Soft Commutation

Soft Commutation

Y. Chéron

CNRS electrotechnology and industrial electronics laboratory,
ENSEEIHT-INP, Toulouse, France

Translated by T. Meynard

CNRS electrotechnology and industrial electronics laboratory,
ENSEEIHT-INP, Toulouse, France

Technical Editor C. Goodman

Department of Electrical Engineering, University of Birmingham, UK

CHAPMAN & HALL
London · Glasgow · New York · Tokyo · Melbourne · Madras

Published by Chapman & Hall, 2–6 Boundary Row, London SE1 8HN

Chapman & Hall, 2–6 Boundary Row, London SE1 8HN, UK

Blackie Academic & Professional, Wester Cleddens Road, Bishopbriggs, Glasgow G64 2NZ, UK

Van Nostrand Reinhold Inc., 115 5th Avenue, New York NY10003, USA

Chapman & Hall Japan, Thomson Publishing Japan, Hirakawacho Nemoto Building, 6F, 1–7–11 Hirakawa-cho, Chiyoda-ku, Tokyo 102, Japan

Chapman & Hall Australia, Thomas Nelson Australia, 102 Dodds Street, South Melbourne, Victoria 3205, Australia

Chapman & Hall India, R. Seshadri, 32 Second Main Road, CIT East, Madras 600 035, India

First English language edition 1992

© 1992 Chapman & Hall

Original French language edition La Commutation Douce – 1989, Technique et Documentation-Lavoisier.

Printed in Great Britain by the University Press, Cambridge

ISBN 0 412 39510 X   0 442 31505 8 (USA)

A catalogue record for this book is available from the British Library

Library of Congress Cataloging-in-Publication data available

Contents

Foreword

FOREWORD

Power Electronics is a part of Electrical Engineering that deals with static converters. These converters enable modification of the form of electrical energy. With this objective in view, they make use of power semiconductors that operate in the switching mode.

Power Electronics has made great strides for two major reasons:

- Electrical energy consumers want energy to be supplied in the form that best fits their application. Encouraged by the capabilities of the converters, they become more and more demanding. They want converters that deliver high grade signals, that inject low noise into the supply and the environment and that respond swiftly. Other requirements are a low price, a small volume and a high reliability.

- Designers are creating more and more techniques to meet such expectations: improvements in existing semiconductors, emergence of new semiconductors, progress in magnetic materials and capacitors, microprocessors allowing reliable and adapting control, improvements in modelling techniques and simulation software...

In the face of such rapid development, the role of scientific publications is very important. They must place new contributions in the context of results previously known, establish links between closely related elements and suggest classifications, ruthlessly eliminate everything that progress makes obsolete and clearly show the principles underlying the current developments. It will be seen how the work of Yvon Cheron fulfills all these requirements.

However, it should first be shown how soft commutation fits into current trends.

One of the major trends in Power Electronics is increasing the frequencies. Higher internal frequencies of operation become possible with faster and faster semiconductors. They result in the passive components of the converters — capacitors, inductors and transformers — becoming smaller thereby reducing the total weight of the equipment. The dynamic performance is also improved.

This frequency elevation is responsible for the growing importance of Pulse Width Modulation on the one hand, and for the use of resonance on the other hand.

Another major trend is the importance given to the reduction of the stresses on the semiconductors and to the conducted and radiated noise generated by the converters.

Resonant converters simultaneously fulfill both these requirements.

In the Pulse Width Modulation Mode, the switches must be turn-on and turn-off controlled. They must allow rapid establishment and blocking of the current. Snubbers must be added in order to limit the current rise at turn on and the voltage step at turn off. However, these snubbers are difficult to implement and introduce losses. They fare what ultimately limits the increase in frequency. In resonant converters, soft commutation can be used. The turn off (or turn on) of the semiconductors occurs inherently[*] with no stress on the semiconductors owing to the natural current (or voltage) zero crossing. This leaves only one turn-off (or turn-on) controlled commutation which can be simply and efficiently softened, by choosing an appropriate converter topology.

Soft commutation allows proper utilization of the semiconductors and thus the attainment of high switching frequencies. A widening of the field of application for soft commutation thus represents undeniable progress; this widening is clearly the most original aspect of Chéron's book.

———————

Mr Cheron, Docteur es Sciences, is a specialist in the study of converter topologies. He is working within a group led by Professor Henri Foch at the Laboratoire d'Electrotechnique et d'Electronique Industrielle of the Institut National Polytechnique de Toulouse.

His book deals with the elements common to all power electronics converters:
- static switches and their classification,
- characterization of electrical sources and the rules governing their interconnection ,
- classification and synthesis of converters,
- duality rules and their consequences,
- the concept of a commutation cell,
- varied types of commutations.

———————

[*] Translator's note: this book introduces the notion of 'inherent commutation' as the broad case of where the switching of a device or a 'commutation cell' (see Chapter 4) occurs due to the zero-crossing of a current or voltage as it varies according to the natural laws governing the circuit operation. In the more special case of, for example, a diode or thyristor in a rectifier configuration this becomes the already familiar case of 'natural commutation'. The full description appears in section 1.2.2.

Using these concepts, he goes on to explain the advantages of soft commutation and of resonant converters.

Besides this, resonant converters are thoroughly reviewed in this book:
- resonant inverters,
- DC/DC converters with a high frequency link.

As far as these converters are concerned, not only are the characteristics determined, but the commutation modes are also compared, the various elements are designed and the modelling of these converters is discussed.

Finally, it is shown how duality — and creative thinking — enables the extension of the field of application of resonance to other types of converters. This extension has given rise to the practical implementations described by the author.

Chéron's book appealed to me for four major reasons:
- It contains a very clear presentation of converters. Such a presentation can save a lot of time for those of us who teach power electronics. I especially appreciated how the concept of a commutation cell is built upon and the systematic use of duality.
- This book is very comprehensive regarding resonant inverters and DC/DC converters using resonant inverters as intermediate stages. In particular, the so-called second-order phenomena which can be very important practically speaking are thoroughly studied.
- Prospective new applications are described. The attractiveness of soft commutation dictates that the extension of its field of application must become a major thrust in the development of power electronics. In Mr Chéron's book, several important results obtained in this field by the group within which he works are presented.
- Finally, this book is clear, carefully written, and pleasant to read. It has a teaching aspect that is, alas, rarely encountered in technical works. For instance, each chapter constitutes a consistent whole with the objectives presented in the introduction and the results obtained summarized in a conclusion.

In my opinion, this book is an original and important contribution to the development of power electronics.

G. SEGUIER
Professor at the Université des Sciences et Techniques de Lille (France)

1
CONSTITUENT PARTS OF STATIC CONVERTERS

1.1 INTRODUCTION

Electrical energy comes in two different forms, either DC or AC, and its characteristics are generally fixed (50Hz, 220/380V mains, DC batteries, etc.). Sometimes the characteristics of this energy can vary over a large range. Examples include alternators or generators driven at a variable speed, photovoltaic cells and wind engines.

However, a lot of applications require specific supplies with fixed or variable characteristics that differ from those of the power sources: 400Hz supplies aboard an aircraft, adjustable DC current for battery chargers, medium frequency (1—100kHz) supplies for induction heating and variable voltage and/or frequency sources for electrical machine drives, for instance.

The electrical energy conversion involves the modification of the form and of the characteristics of the electrical energy delivered by the supply in order to match the application.

The scope of application is very wide and impinges on almost all areas of industry: battery chargers, uninterruptible power supplies, soldering, heating, ultrasonic treatment, electrolysis, VHV generators for X-ray valves or radar, speed control of electrical machines, active filtering, network compensation, etc. Powers controlled range from a few watts to a few megawatts.

These conversions, formerly achieved by means of electromechanical converters (mainly rotating machines), are nowadays achieved by **static converters** that are less heavy and more effective.

A static converter is composed of a set of electrical components building a meshed network that acts as a linking, adapting or transforming stage between a generator and a load.

An ideal static converter enables the controll of the power flowing from the generator to the load with a 100% efficiency. No loss should theoretically occur in the converter. The basic constituents are of two types:

- Non-linear elements which are, most of the time, semiconductor electronic switches,
- Linear reactive elements: capacitors, inductors (and transformers). It should be noted that these reactive components play a part in energy storage and transformation but also perform voltage and current filtering. They generally represent an important part of the size, weight and cost of the equipments.

1

The static converters using semiconductors form a very wide and varied group because of the many types of conversions to be achieved (AC/DC, variable frequency/fixed frequency, etc.), and because of the numerous types of semiconductors with different features (diodes, thyristors, bipolar transistors, MOSFETs, GTO, triacs and so on). The different types of conversion and the usual names of the corresponding converters are given in Fig. 1.1.

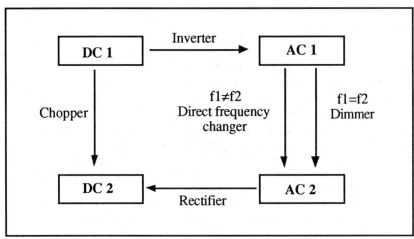

Fig. 1.1 Main types of converters

In this chapter a certain number of concepts specific to the static converter field such as the concept of a switch or the concept of a source will be recalled. Other definitions concerning the static and dynamic characteristics of a switch and the inherent or controlled commutation of this switch will also be given.

The switches and the sources are the basic constituents of static converters; they are connected together according to very simple rules that constitute the fundamental principles of energy conversion and from which the converter topologies are derived.

1.2 THE CONCEPT OF A SWITCH

Static converters are electrical networks mainly composed of semiconductor devices operating in the switch mode (**switches**) and allow, through proper sequential operation of these devices, an energy transfer between a generator and a load bearing different electrical characteristics.

In order to maximize the efficiency of the converter the losses in the switches must be minimized. With this aim in view, the switches must exhibit a voltage drop (or an ON resistance) as low as possible when turned on, and a negligible leakage current in the OFF state.

1.2.1 Static characteristics

Taken as a dipole with the **load** sign convention (Fig. 1.2), the **static characteristic** $I_K(V_K)$ that represents the operating points of a switch is made up of two branches totally located in quadrants 1 and 3 such that $V_K*I_K>0$. One of these branches is very close to the I_K axis (ON state) and the other is very close to the V_K axis (OFF state), each of these branches possibly being unidirectional. Idealizing this switch, the static characteristic can be assimilated to the half-axis to which it is close.

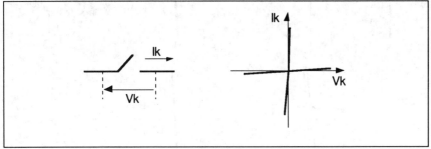

Fig. 1.2 Static characteristic of a switch

In this representation, except for the trite cases of the short circuit and of the open circuit that correspond respectively to a switch always ON and to a switch always OFF, any switch that really operates as such has a static characteristic consisting of at least two orthogonal half-axes (or segments).

The static characteristic that is an intrinsic feature of a switch thereby reduces to a certain number of segments in the $I_K(V_K)$ plane.:

• **Two segments** if the switch is unidirectional for current and for voltage. Two two-segment characteristics must be distinguished: in the first case current I_K and voltage V_K always have like signs; in the second case current I_K and voltage V_K always have opposite signs. The switches bearing such characteristics are respectively called **T** and **D** (Fig. 1.3).

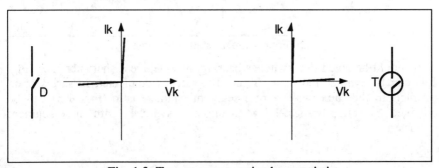

Fig. 1.3 Two-segment static characteristic

3

• **Three segments** when either the current or the voltage is bidirectional while the other is unidirectional. So there are two types of three-segment static characteristics (Fig. 1.4).

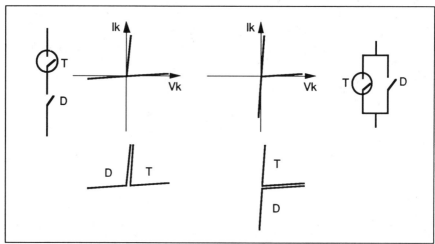

Fig. 1.4 Three-segment static characteristic

• **Four segments** when voltage and current are bidirectional. There is only one such type of static characteristic (Fig 1.5).

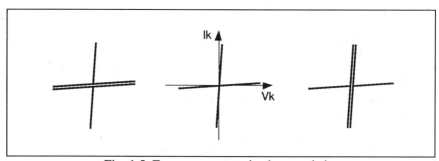

Fig. 1.5 Four-segment static characteristic

It should be noted that switches bearing three-segment characteristics (Fig. 1.4) can be synthesized using **T** and **D** switches connected in series or in parallel. In the same way, a four-segment characteristic (Fig. 1.5) can be obtained by series or parallel connection of switches with three-segment characteristics.

1.2.2 Dynamic characteristic

The voltage-current static characteristic of a switch is, however, inadequate to describe the dynamic properties, i.e. the way the off switch turns on and conversely. The **commutation dynamic characteristic** is the trajectory described by the point of operation of the switch during its commutation, to go from one half-axis to the perpendicular half-axis. A switch being either ON or OFF, there are two commutation dynamic characteristics corresponding to the turn on and the turn off, which will be grouped under the global term of **dynamic characteristic**.

Unlike the static characteristic, the dynamic characteristic is not an intrinsic property of the switch but depends on the constraints imposed by the external circuit.

Neglecting second order phenomena, and taking account of the dissipative nature of the switch, the dynamic characteristic can only be located in those quadrants where $V_K*I_K>0$.

(a) Inherent commutation

From the former remarks it can already be inferred that the commutation dynamic characteristic at turn on or at turn off of switch **D** necessarily merges into the static characteristic. So, the commutation of this switch, turning off when the current flowing through it vanishes and turning on when the voltage applied across its terminals reaches zero, is fully dependent on the evolution of the electrical quantities in the external circuit. This commutation being, by nature, identifiable with that of a PN junction, switch **D** is in most cases a diode and this type of commutation is referred to as **inherent**. Using any other controlled semiconductor leads to a complex control that must be synchronized to the electrical quantities of the external circuit.

Such a commutation is achieved with minimum losses since the operating point moves along the axes.

(b) Controlled commutation

By virtue of the former remarks, switch T cannot commutate inherently since I_K and V_K always have like signs. Nevertheless the internal resistance of this switch can change from a very low value to a very high value at turn off (and conversely at turn on), and this is independent of the evolution of the electrical quantities imposed on switch **T** by the external circuit. So this switch has, in addition to its two main terminals, a control terminal on which it is possible to act in order to provoke a quasi-instantaneous change of state. Such a commutation will be referred to as a **controlled commutation**.

It should be noted that in a controlled commutation, the switch imposes its state on the external circuit. Under such circumstances, the element can undergo severe stresses that depend on its dynamic characteristic. If the switching time is long and the operating frequency is high, the ohmic losses can be significant.

5

1.3 CLASSIFICATION OF SWITCHES

Switches used in static conversion can be classified by their static characteristics with two, three or four segments and by the type of commutation (inherent or controlled) at turn on and at turn off.

Let us recall that a controlled commutation can occur only in the first and third quadrants while an inherent commutation can occur only in the second and fourth quadrants.

1.3.1 Two-segment switches

Putting aside the open circuit and the short circuit, two switches with two-segment characteristics can be distinguished (Fig. 1.6):

• The first of these switches has the static characteristic of switch **D** and its turn on and turn off commutations are inherent.(Fig. 1.6a). This switch is a diode and it is symbolized as such.

•The second of these switches bears the static characteristics of switch **T** and its turn on and turn off commutations are controlled (Fig. 1.6b). This switch represents the modern power semiconductors: MOS and bipolar transistors, Darlingtons, Gate Turn Off thyristors and the newer components such as COMFETs, GEMFETs or similar devices.

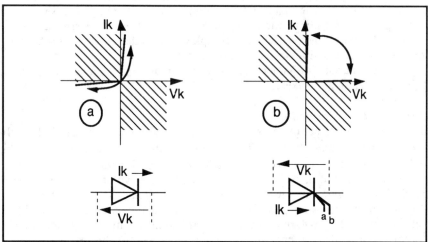

Fig. 1.6 Two-segment switches

This switch will be symbolized by separating the turn-on and turn-off control as shown on Fig. 1.6b.

6

Any other switch that would have the same static characteristic as transistor **T**, but that would only have one controlled commutation (either turn on or turn off), could not be used on its own. In effect, an inherent commutation must compulsorily be added to the controlled commutation and this inherent commutation can be obtained only if the static characteristic includes two segments with opposite signs. It shows that such a switch must necessarily be associated with a series or parallel diode to obtain finally a three-segment switch.

Thus, only two two-segment switches can be used directly.

1.3.2 Three-segment switches.

These switches can be divided into two groups depending on whether they are bidirectional for current and unidirectional for voltage (Fig. 1.7), or unidirectional for current and bidirectional for voltage (Fig. 1.8). In each of these groups all switches have the same static characteristic and differ only in their commutation mechanisms.

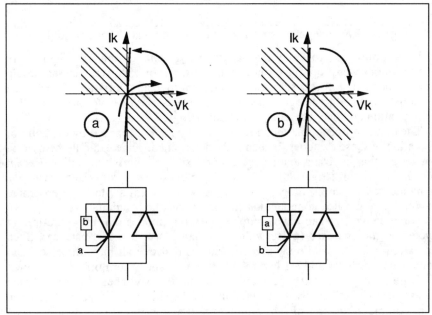

Fig. 1.7 Three-segment switches with bidirectional current capability

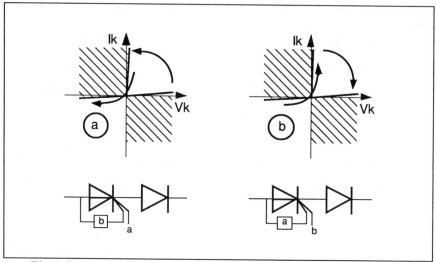

Fig. 1.8 Three-segment switches with bidirectional voltage capabilty

It is important to note that a three-segment switch in which both commutations were controlled (turn on and turn off) or both were inherent would never utilize the three segments of its static characteristic. Under such circumstances, a three-segment switch must necessarily have one controlled commutation and one inherent commutation (Figs. 1.7 and 1.8).

The cycle of operation, which represents the locus described by the point of operation of these switches, is then fully determined. They can only be used in converters that impose a unique cycle on the switches during operation. Except for the thyristor (Fig.1.8a), all these switches are synthesized switches, implemented by connecting a diode to a two-segment switch, the ensemble having one controlled commutation and one automatic commutation.

Some converters can have complex operating modes that impose different cycles on the switches (pulse width modulated inverters for instance). Such operation can necessitate the use of switches that overall have three-segment static characteristics and are controlled at turn on and at turn off. But whatever the operating mode, all the capabilities of the switches are never used simultaneously, especially the reversibility of the current or of the voltage when both commutations are of the same nature, i.e. controlled or inherent. (Then these switches behave as true two-segment switches.)

1.3.3 Four-segment switches

All four-segment switches have the same static characteristic. They differ only by their commutation modes that can, a priori, differ in quadrants 1 and 3. So, **six four-segment switches** can be distinguished.

These switches are used mainly in direct frequency changers; in practice they are made up of two three-segment switches series or parallel connected.

1.3.4 Switch representation

In the following, to simplify the representation of converter circuits, the two-segment and three-segment switches are symbolized as shown on Fig. 1.9. The automatic control that ensures the inherent commutation of the switch is not shown.

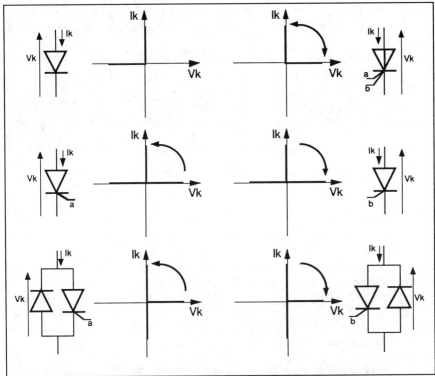

Fig. 1.9 Symbols for the two- and three-segment switches

1.4 CHARACTERIZATION OF SOURCES

1.4.1 The concept of sources

The current (or the voltage) that characterizes a generator or a load is called **DC** if it is **unidirectional**; as a first approximation, it can be taken as constant. The current (or voltage) is called **AC** if it is periodic and has an average value equal to zero; as a first approximation it can be taken as sinusoidal.

A generator or load is **voltage-reversible** if the voltage across its terminals can change sign. In the same way, a generator or a load is **current-reversible** if the current flowing through it can reverse.

In the case of a change of direction of the energy flow, i.e. a change in the sign of the power, generators and loads can exchange their functions briefly (**instantaneous reversibility**) or persistently (**permanent reversibility**).

So the input/output of a converter can be characterized as voltage or current sources (generators or loads), either DC or AC, current-reversible and/or voltage-reversible. Any electrical dipole connected to the input or to the output of a converter can thus be represented by one of the eight types shown in Fig. 1.10.

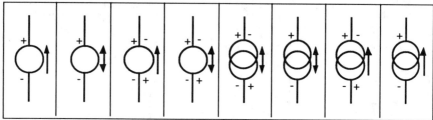

Fig. 1.10 Various types of sources involved in power converters

1.4.2 Nature of sources

The principle of operation of a converter is based on the switch mode action of its constituent switches. Commutations of these switches generate very fast current and/or voltage transients so that the transient behaviour of the sources is of crucial importance for the converter.

The transient behaviour of a source is characterized by its ability or inability to withstand steps in the voltage across its terminals or in the current flowing through it, **these steps being generated by the external circuit**. Two types of sources can be distinguished:

• **Voltage sources** of which the voltage cannot undergo a step. The most representative example is the capacitor since an instantaneous change of voltage would correspond to an instantaneous change of its charge which would require an infinite current.

• **Current sources** of which the current cannot undergo a step. The most representative example is the inductor since an instantaneous change in current would correspond to an instantaneous change in its flux which would require an infinite voltage.

It should be noted that a square wave voltage generator (respectively current generator) is indeed a voltage source (respectively current source) as defined above since the voltage steps (respectively current steps) are not caused by the external circuit.

Under such circumstances, we come to define the notion of instantaneous impedance of a source as the limit of the source impedance when the Laplace operator s tends to infinity. Theoretically this instantaneous impedance can be zero, finite or infinite.

A source is referred to as a **voltage source** when its instantaneous impedance is zero, while a source is called a **current source** if its instantaneous impedance is infinite. The particular case of the non-zero finite instantaneous impedance is not considered in this presentation for two reasons:

- Except for special cases (heating for example), the resistors are parasitic elements that can be neglected without restricting the scope of the work.
- A resistor generally has some inductive impedance.

It should be noted that connecting a series inductor[*] with an appropriate value to a voltage source (that is a dipole with a zero instantaneous impedance) turns the voltage source into a current source. In the same way, connecting a parallel capacitor of appropriate value to a current source (dipole with an infinite instantaneous impedance) turns the current source into a voltage source.

These inductive or capacitive elements connected in parallel or in series with the source are elements that can temporarily store energy. Consequently, if an inductor connected to a voltage source turns it into a current source it is important to determine the current reversibility of this current source.

Obviously, the current source obtained by connecting an inductor in series with a voltage source keeps the same current reversibility as this voltage source. The inductor acts as a buffer absorbing the voltage differences. Consequently, the current source obtained by connecting a series inductor to a voltage source is voltage reversible. When the voltage source itself is voltage reversible there is no particular problem. On the other hand, if the voltage source is not voltage reversible, the current source obtained by connecting a series inductor to the voltage source is only instantaneously reversible with respect to voltage.

The former results can easily be transposed to the voltage source obtained by parallel connection of a capacitor to a current source. The voltage source obtained bears the same voltage reversibility as the current source, and is current reversible, but this reversibility is only instantaneous if the current source is not current reversible.

---

[*] Inductors and capacitors can, in some cases, be replaced with dipoles that also have, respectively, an infinite or zero instantaneous impedance, for example a series LC network or a parallel LC network.

1.4.3 Examples

A set of ideal batteries behaves as a load during charging and as a generator during discharging; such a source is called a DC voltage source, current reversible but not voltage reversible. Nevertheless, because of the inductance of the connecting cables, this battery can sometimes be taken as a current source instantaneously voltage reversible and permanently current reversible.

The armature circuit of a DC machine can be represented by an e.m.f. in series with a resistor and an inductor. The polarity of the e.m.f. depends on the direction of rotation and the sign of the current depends on that of the torque (either motoring or braking). The DC machine in the generator mode, generally regarded as a voltage generator by users, must be regarded here as a current generator because of the inductance of its armature winding. Hence this current source is a DC source and is both current and voltage reversible.

In practice the identification of a real generator or of a real load with a voltage or current source is seldom obvious because of their non-ideal characteristics. That is the reason why the nature of the source is often reinforced by the addition of a parallel capacitor in the case of voltage sources and by the addition of a series inductor in the case of current sources.

1.5 BASIC PRINCIPLES OF STATIC CONVERTERS

A static converter consists mainly of switches. The control of turn on and turn off of these switches according to specified cycles enables periodic modification of the direct interconnection of the terminals of the sources connected to the input and the output of the converter thus allowing control of the power flow between the two sources. These sources are respectively called the **input source** and the **output source**.

The input source and output source interconnection laws rule the converter design and can be stated very simply:

- A voltage source should never be short-circuited but it can be open-circuited.
- A current source should never be open-circuited but it can be short-circuited.

From the two former rules it can readily be inferred that the switches cannot establish straight connection between two voltage sources, which define the voltages across their terminals, or between two current sources, which define the current they carry.

Nevertheless, if one of the voltage sources (respectively one of the the current sources) does not impose the voltage across its terminals directly (respectively the current), which is for instance the case of the RC parallel network (respectively the LC series network), it is possible to connect these two voltage sources (respectively current sources) provided certain precautions are taken:

• In the case of two voltage sources, turn on of the switch can only occur when the two sources have the same values, that is to say at the zero crossing of the voltage across the switch. The turn on must then be inherent (since it depends on the external circuit) and turn off can be controlled at any time.

• In the case of two current sources, turn off of the switch can only occur when the two current sources reach the same value, that is to say when the current in the switch vanishes. So turn off is inherent and turn on can be controlled at any time.

This reasoning results in definitions of the commutation mechanisms of the switch(es) that must interconnect two sources of the same type.

In practice the converters in which two voltage sources (or two current sources) are interconnected are very exceptional. As far as we know there is but one such type of converter that is used commonly. It is the diode rectifier feeding a load connected to a parallel capacitor and supplied directly, or through a transformer, by the mains which can be regarded here as an AC voltage source. Such configurations are also met when the switches are equipped with turn-on snubbers (in rectifiers) or with turn-off snubbers (in inverters with dual-thyristors: see Chapter 2).

Except in the case of special precautions concerning the dynamic characteristics of the switches, these switches can establish direct connections only between two sources of different types.

In all of the following presentation, only voltage-current converters (and the reverse) will be studied.

1.6 STRUCTURES OF STATIC CONVERTERS

The structures of static converters are logically derived from the nature of the generators and the loads to which they are connected. The structures also depend on the desired energy transfer reversibility and on the mode of control of these energy transfers.

The structure of a converter is determined fully when its topology and the nature of its constituent parts are known.

An electric energy conversion can be realized by means of different structures with one or several intermediate conversion stages. When this conversion is achieved without any intermediate stage temporarily storing some energy, the conversion is called **direct conversion** and it is achieved by a **direct converter**. On the other hand, when this conversion makes use of one or more stages able to store energy temporarily, the conversion is called **indirect** and it is achieved by an **indirect converter**.

1.6.1 Direct converters

A direct converter is an electrical network composed only of switches and is thereby totally unable to store energy. In such a converter, energy is directly transferred from the input to the output and, provided the losses in the converter can be neglected, the input power is at any time equal to the output power.

Under such circumstances, if the input source of a direct converter is a voltage source, its output obviously behaves as a voltage source as well and can only be connected to a current source. (The phenomena of discontinuous conduction are not taken into account in this presentation.)

Similar reasoning can also be applied to determine the nature of the output of a direct converter when a current source is connected at the input, or to deduce the nature of the input source from the output source. The following fundamental rule can then be deduced: **'The input (respectively the output) of a direct converter is of the same type as the output source (respectively the input source.)'**

Furthermore, we will use the name **elementary direct converter** for any converter that employs the simplest structure to achieve the desired conversion, that is to say, the structure that uses the minimum amount of switches. A direct converter can then be obtained by grouping several elementary direct converters provided no other element is added.

Such an association may prove necessary when the electrical energy must be present in a form that differs from that imposed by the supply and by the load in order to fulfill some requirements of the load specification (for instance, in a DC-DC converter, an intermediate AC stage is necessary when galvanic isolation is required).

(a) Synthesis of direct converters

Taking account of the interconnection rules stated above, the different possible connections between a dipole voltage source and a dipole current source are shown in Fig. 1.11. The more general structure that achieves these connections is the four switches bridge structure (Fig. 1.12). When some of these connections are not necessary, it is possible to degenerate the bridge structure into structures using fewer switches.

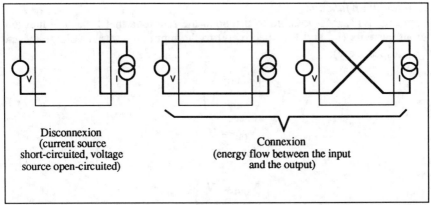

Fig. 1.11 Connexions between a voltage source and a current source in a direct converter

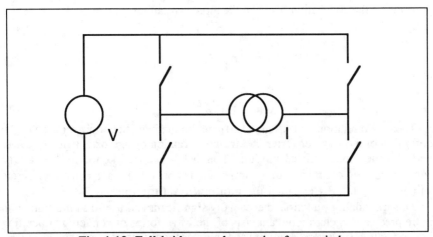

Fig. 1.12 Full-bridge topology using four switches

To characterize a switch, we have to determine its static and dynamic characteristics. To determine the static characteristic of a switch, it is sufficient to determine the sign of the current through it in the ON state and the sign of the voltage across it in the OFF state. Establishing the static characteristics thus requires little information, since only the reversibility of the sources effectively used in the planned converter need be known. By contrast, finding out the dynamic characteristics calls for a thorough knowledge of the various operating phases, that is to say the desired operation of the converter.

The synthesis of direct converters based on the principles stated above has already been thoroughly studied [24,25].

(b) Modulation function

From Fig. 1.11, it is possible to establish the relations that link the input and output currents (I_{in} and I, respectively) to the input and output voltage (V and V_{out}, respectively):

$$\left\{ \begin{matrix} V_{out} = 0 \\ I_{out} = 0 \end{matrix} \right\} \tag{1.1}$$

$$\left\{ \begin{matrix} V_{out} = V \\ I_{out} = I \end{matrix} \right\} \tag{1.2}$$

$$\left\{ \begin{matrix} V_{out} = -V \\ I_{out} = -I \end{matrix} \right\} \tag{1.3}$$

All these equations can be grouped by using a coefficient F_m that only depends on the connection between the input and output sources.

$$V_{out} = F_m V \tag{1.4}$$

$$I = F_m I_{in} \tag{1.5}$$

where

$$F_m = -1, 0, +1 \tag{1.6}$$

These two relations ((1.4) and (1.5)) are the mathematical transcription of the definition of a direct converter, i.e. the conservation of the instantaneous power between the input and the output. Though F_m can be likened to the voltage conversion ratio of an ideal transformer, F_m is not constant and can take discrete values -1, 0, +1. F_m is called the **modulation function**.

This modulation function is especially significant and useful in the study and more precisely the harmonic analysis of the wave forms of converters operating by modulation (inverters, rectifiers and more generally direct frequency changers) [26~29].

It should be noted that this modulation function can be used only if the converter operation **exclusively depends on the switch control**. It cannot be used in converters in which operation depends on the input and output source imperfections. We think here mainly of dimmers, converters operating in the discontinuous conduction mode, or else of those in which the commutations cannot be taken as instantaneous (overlap phenomenon).

1.6.2 Indirect converters

An indirect converter is generally made of an association of direct converters and reactive elements such as inductors and capacitors. The electrical energy flowing between the input and output sources crosses these elements without any active power being lost.

The essential principle of an indirect converter is the cascade connection of at least two direct converters, the two of them being connected through an intermediate stage only comprising reactive elements. In particular an indirect converter allows independent control of the different quantities in each direct converter.

Indirect energy conversion is mainly used when the two sources to be connected by means of the converter are of the same type, an intermediate stage of a different type then being absolutely necessary. This buffer stage is a voltage source if the energy transfer involves two current sources, and it is a current source if energy must be transferred between two voltage sources. In practice it can also be attractive in order to achieve simply a complex conversion (utilization of a DC buffer stage in AC/AC conversion).

Some indirect converters can also be simplified when they use special controllers. A notable case of this is DC/DC converters with inductive or capacitive storage.

The synthesis of an indirect converter resolves, in fact, to the synthesis of the direct converters that it is made of, the definition and placing of the reactive elements and the connection of these reactive elements with the direct converters following the rules for the interconnection of sources. This synthesis is possible only if the nature and reversibility of the input and output sources of each direct converter are precisely defined [24,30,31].

The nature of the input source (respectively the output source) of a direct converter being fully imposed by the network connected at the output (respectively the input) of this converter, the input and output sources of each direct converter used in an indirect energy conversion are exactly known. The problem of the indirect converter synthesis is thus solved.

1.7 CONCLUSION

In this introductory chapter, we have mainly:
- characterized the basic constituent parts of the static converters (switches and sources),
- reviewed the few essential laws that rule their operation,
- specified a few definitions (reversibility, modulation function).

Though our main purpose was to specify the vocabulary used in the following text, mastering these concepts is also necessary for a global and synthetic approach to static conversion.

These concepts allow us to understand the function implemented by a converter and enable us to determine its structure, regardless of the often complex second order phenomena.

From an educational viewpoint, they are essential and have therefore already been intensively used in power electronics courses.

2

THE CONCEPT OF DUALITY IN STATIC CONVERTERS

2.1 INTRODUCTION

Presenting an approach to the synthesis of the very numerous power electronics circuits naturally leads to an attempt to establish relations between these circuits.

The constituent parts of the converters are mainly switches and reactive elements. These switches are either ON ($V_K=0$) or OFF ($I_K=0$), uni- or bidirectional for voltage and/or for current, and controlled at turn on and/or at turn off. The reactive elements are inductors (either coupled or not) and capacitors able to store some energy in the form of current and voltage, respectively. In other respects, these converters control the energy flowing between sources characterized as current or voltage sources.

Among these elements, there is an obvious one-to-one relation between elements with given properties for voltage (respectively current) and elements with the same properties for current (respectively voltage). This relation, which indeed is not new, is referred to as **duality**.

After a brief reminder of the basic concepts of duality in graphs and circuits [32,33] (the reader can consult the specific references for further information), we will show in this chapter how the concept can be extended to power electronics circuits [20].

As these circuits involve switches, it is necessary to define the dual elements of these switches. In particular, this leads to the definition of a 'new' switch: the 'dual-thyristor'. Finally, we try to extend the application of duality rules to more complex converters, especially the polyphase converters that cannot be drawn on a plane without any branches crossing.

2.2 REVIEW OF DUALITY APPLIED TO GRAPHS AND CIRCUITS

Duality enables definition of pairs of dual (or corresponding) electrical quantities such as:

- Current — Voltage
- Resistance — Conductance
- Inductance — Capacitance
- Impedance — Admittance
- Flux — Quantity of electricity

It should be noted that both electrical power and electrical energy correspond with themselves.

Duality only applies to graphs and circuits that can be drawn on a plane without any branch crossover (this limitation is especially significant in the case of polyphase converters).

Two graphs are dual if:
- there is a one-to-one relation between the circuit loops of one of the graphs (including the external loop) and the nodes of the other graph,
- each branch 'B' common to two loops of one of the graphs can be related, in the other graph, to a branch 'B*' that links two nodes related to these two loops.

Two electrical circuits are dual when:
- their graphs are dual,
- the elements of the dual branches B and B* of these two graphs are dual.

A classic example of two dual circuits is given in Fig. 2.1.

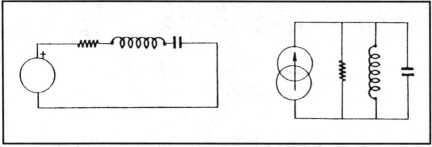

Fig. 2.1 Example of circuits which are the dual of each other.

The equations describing the operation of these two circuits can be written in the same symbolic form:

$$\left\{ \begin{array}{l} V(s) = R.I_V(s) + L.s.I_V(s) + \dfrac{1}{Cs} . I_V(s) \\[2mm] I(s) = G.V_I(s) + C'.s.V_I(s) + \dfrac{1}{L's} . V_I(s) \end{array} \right\} \tag{2.1}$$

where V, R, L, C and I_V are respectively the values of the voltage source, the resistance, the inductance, the capacitance and the current flowing in the voltage source in the series circuit, and I, G, C', L' and V_I are respectively the values of the current source, the resistance, the inductance, the capacitance and the voltage across the current source in the parallel circuit.

The dual circuit of a given circuit is obtained by marking a point in each loop of the circuit and one point outside the circuit, this point being related to the external loop (Fig. 2.2). Each of these points represents a node of the dual circuit. Between any two points and for each branch common to the two loops that surround these points, we draw a branch with the corresponding element of the element on the branch common to the said loops.

A simple example of obtaining a dual circuit is given in Fig. 2.2.

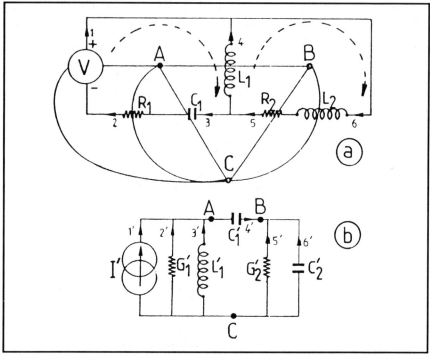

Fig. 2.2 Obtaining the dual of a circuit

After obtaining the dual circuit, it is necessary to define the sign of the dual quantities such as the sign of current source I' in Fig. 2.2b. With this aim in view, we have to orient the branches of the original circuit and to derive the orientation of the branches of the dual circuit. Taking account of the orientation chosen, the loops rule applied to the circuit in Fig. 2.2a reads:

$$V_1 + V_2 + V_3 - V_4 \qquad = 0 \qquad (2.2)$$

$$V_4 + V_5 + V_6 \qquad = 0 \qquad (2.3)$$

$$V_1 + V_2 + V_3 + V_5 + V_6 = 0 \qquad (2.4)$$

in which V_i is the voltage across the branch labelled i.
At node A, the nodes rule reads, by duality:

$$I_1 + I_2 + I_3 - I_4 = 0 \qquad (2.5)$$

and at nodes B and C:

$$I_4 + I_5 + I_6 = 0 \qquad (2.6)$$

$$I_1 + I_2 + I_3 + I_5 + I_6 = 0 \qquad (2.7)$$

The orientation of the circuit branches in Fig. 2.2b is directly derived from expressions (2.5), (2.6) and (2.7).

This determination of the sign of the dual quantities is essential when the circuit involves unidirectional elements.

2.3 DUAL STATIC SWITCHES

The notion of duality can be extended to static converters which are basically electrical circuits. Nevertheless, to be totally applicable, it requires the determination of the dual elements of the switches.

Some authors [34,35] have circumvented this difficulty by decomposing the converter operation into several phases. The ON switches (respectively OFF switches) are replaced with short circuits (respectively open circuits) in the circuits corresponding to each of these phases. Applying duality rules to each of these circuits gives the circuits corresponding to each phase of operation of the dual converter that can finally be identified.

In this form, this method suffers from an important deficiency since only the operation between commutations, which only determines the static characteristics of the switches, is taken into account. The commutations of the switches (dynamic characteristic) are thus hidden.

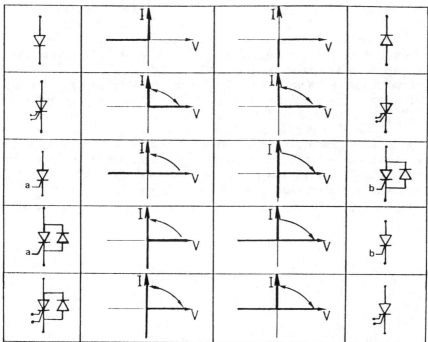

Fig. 2.3 Duality correspondence between various types of switches

Figure 2.3 shows the different two-segment and three-segment switches and shows the duality relations existing between them. Obviously the diode and the components of the 'transistor type', which are two-segment switches, are their own dual.

As far as the three-segment switches are concerned, applying duality to a switch with turn-on and turn-off control leads to a switch that is easily identified. By contrast, the turn-on controlled switches with inherent turn-off, that is to say the switches of the 'thyristor type' with or without an antiparallel diode, have no discrete dual. Nevertheless, these dual switches can be synthesized from a 'new' type of switch connected to a series diode or not and that bears the dual characteristics of those of a thyristor. This switch is referred to as a **dual-thyristor** in all the following text.

2.4 THE DUAL-THYRISTOR

From the characteristics of the thyristor it is possible to apply duality rules in order to define the characteristics of the dual-thyristor without making any assumption pertaining to its constituent parts [19,20].

The dual-thyristor turns on if its control allows it **and** if the voltage across its terminals is zero, it can be turned off by a command if the current flowing through it is positive. This turn-off control is ineffective when the current is negative, exactly like the turn-on control of a reverse-biased thyristor.

The turn-off losses are zero in a thyristor since this turn off occurs under zero current (inherent commutation). Similarly, in a dual-thyristor the turn-on losses are zero since this commutation occurs under zero voltage.

A series inductor connected to a thyristor protects the thyristor against high dI/dt at turn on, thus limiting the switching losses at turn on. From this, it can readily be inferred that a parallel capacitor connected to a dual-thyristor protects the switch against high dV/dt at turn off, thus limiting the turn-off losses.

An attractive property of the dual-thyristor (attractive because of the safety of operation it induces) is the **interruption**. In effect, if a thyristor turns on when an overvoltage is applied across its terminals ($V=V_M$), it means that an overcurrent in a dual-thyristor causes its turn off ($I=I_M$).

The properties of the thyristor and of the dual-thyristor are grouped in Table 2.1.

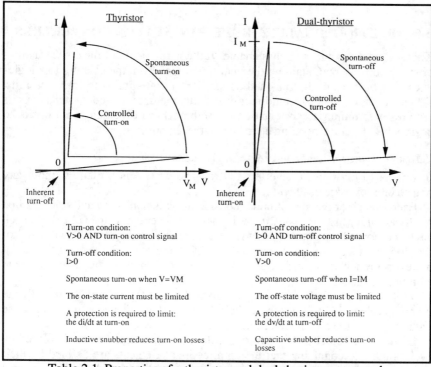

Turn-on condition:
V>0 AND turn-on control signal

Turn-off condition:
I>0

Spontaneous turn-on when V=VM

The on-state current must be limited

A protection is required to limit:
the di/dt at turn-on

Inductive snubber reduces turn-on losses

Turn-off condition:
I>0 AND turn-off control signal

Turn-on condition:
V>0

Spontaneous turn-off when I=IM

The off-state voltage must be limited

A protection is required to limit:
the dv/dt at turn-off

Capacitive snubber reduces turn-on losses

Table 2.1 Properties of a thyristor and dual-thyristor compared

In practice, the dual-thyristor is made from the usual components (transistors and diodes for instance) equipped with special control circuits. It is symbolized as shown in Fig. 2.4.

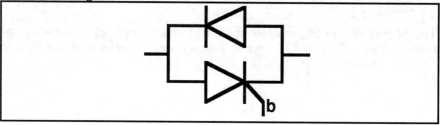

Fig. 2.4 Symbol for the dual-thyristor

2.5 APPLYING DUALITY RULES IN STATIC CONVERTERS

Knowledge of the dual switches enables full determination of the dual structures of static converters. Obtaining the dual topologies from the existing topologies is carried out according to a method similar to that applied to linear circuits. However, the presence of unidirectional components in the converter's circuits requires determining the orientation of the branches of the dual circuit prior to positioning the corresponding elements on these branches.

2.5.1 Obtaining the dual converter of a given converter

In this part only those converters that can be drawn on a plane without any branch crossover are dealt with.

Each branch of the circuit under consideration is numbered and oriented in a totally arbitrary fashion. The directed graph for the circuit is then drawn. In each adjacent loop of this graph a point is marked corresponding to the dual node for that loop and also a point is marked outside the circuit. Between each pair of nodes, a branch is inserted for each element common to the two loops that surround the nodes.

With a direction of traversal fixed for a loop in the original graph, all the branches oriented in this way have their dual branches oriented in the same sense with respect to the dual node of that loop.

Having also directed the dual graph, the corresponding element is placed on each branch. Recognizing that the unidirectional elements are defined by their voltage-current characteristics, each branch oriented by the sense of the current (voltage) of the characteristic of the element it carries has a dual branch oriented by the sense of the voltage (current) of the characteristic of the corresponding element.

2.5.2 Examples

The method given here to obtain the dual converter of a given converter is totally general. It is applied here to find out the dual of a thyristor forced commutated chopper (Fig. 2.5).

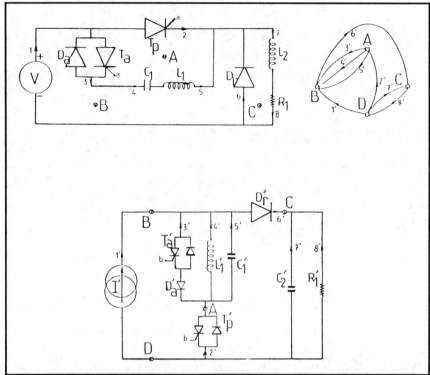

Fig. 2.5 Obtaining the dual converter of a forced commutated thyristor chopper

The bridge structure is invariant by duality. We just have to replace each element by its dual and to exchange K'_2 and K'_3 (Fig. 2.6).

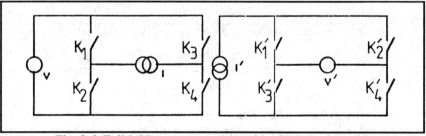

Fig. 2.6 Full-bridge converter (a) and its dual topology (b)

In the examples of Figs 2.7, 2.8 and 2.9, the development of the dual circuits is based on the preceding remark.

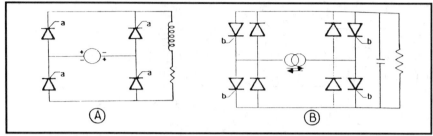

Fig. 2.7 Thyristor voltage rectifier (a) and its dual topology (b)

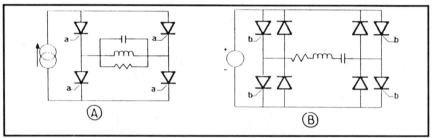

Fig. 2.8 Thyristor parallel current-source inverter (a) and its dual topology (b)

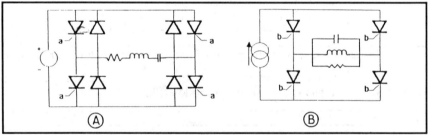

Fig. 2.9 Thyristor series voltage-source inverter (a) and its dual topology (b)

It is also quite easy to demonstrate that the buck converter is the dual of the boost converter and that the capacitive storage chopper (Cük converter) is the dual of the inductive storage chopper (buck-boost).

N.B. An important phenomenon in DC/DC converters is the discontinuous conduction mode in which the inductor current reaches zero before the end of the restoring phase. This phenomenon has a dual in the capacitive storage converters and it consists of a total discharge of the capacitor during the energy restoring phase.

2.6 EXTENSION TO COMPLEX STRUCTURES

Up to now, duality rules have been applied fully to those static converters circuits that can be drawn on a plane without any branch crossover and without any transformer. These two restrictions prevent us from generalizing the duality, a problem which can be overcome by focusing on the function of the converter instead of its circuit.

A transformer is, in the simplest case, made up of two coupled inductors. The possibility of coupling inductors has no dual for capacitors. Nevertheless, as far as the transformer effect is concerned, that is to say a voltage and current transformation such that input and output powers are equal, it is apparent that the dual of an ideal transformer is an ideal transformer also.

Those converters that cannot be drawn on a plane without branch crossover are mainly the DC/AC polyphase converters and, more generally, the direct frequency changers.

These direct frequency changers are circuits that control the energy flowing between a polyphase voltage source and a polyphase current source. They consist basically of a matrix of switches capable of interconnecting each phase of one of the systems to any phase of the other.

It should readily be appreciated that decomposing the operation of the converter into different modes, applying duality rules to the circuits corresponding to the different modes, and finally identifying the dual converter (using the dual circuits obtained at the previous step), is a method that can be used as soon as each of the said circuits can be drawn on a plane. Nevertheless, to obtain real dual converters, it is obvious that even the switches that compose them must also have dual dynamic characteristics.

In the interests of brevity, only the problem of obtaining the dual converter of the m-phase rectifier (or Graetz bridge) with m > 2 is discussed. It is a particular case of a direct frequency changer.

In the general case, a Graetz bridge is a static device (Fig. 2.10a) that enables the connection across a DC current source of any of the line-to-line voltages of a star-connected polyphase source, either balanced or not (delta-connection implies that the system is balanced).

Applying duality, it is shown that a dual Graetz bridge is a static device (Fig. 2.10b) that enables the imposition through a DC voltage source of any of the line currents of a delta-connected polyphase current source, either balanced or not (star-connection without neutral implies that the system is balanced).

The m-phase Graetz bridge is, like its dual, composed of a 2m-switches matrix. These switches have dual static and dynamic characteristics.

Disregarding the commutation phenomena, avoiding voltage source short-circuits and current source open-circuits means that only one switch can be on at any time in each column of the matrix in Fig. 2.10a, these two switches being possibly on the same line which would correspond to shorting the load (free-wheeling phase). This shows that the waveform of the voltage across the current source can be taken as the difference of two m-pulse voltage waveforms.

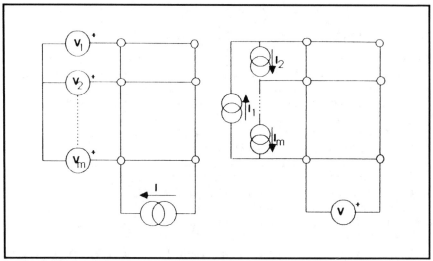

Fig. 2.10 Applying duality rules to a polyphase converter (Graetz rectifier)

For the same reasons, one and only one switch can be ON at any instant in each line of the matrix of Fig. 2.10b, these m ON switches being possibly in the same column when no current will flow in the output source. The waveform of the current flowing in the voltage source is the sum of m two-pulse current waves.

This analysis shows a basic difference between the operation of these two converters having dual functions and using almost identical circuits neither of which can be drawn on a plane. Notice that the two methods of operation are exactly dual in the particular case of m=2; in this case, the circuit can be drawn on a plane.

These dual circuits are drawn in more detail in Fig. 2.11 for the case of a three-phase balanced AC source (m=3).

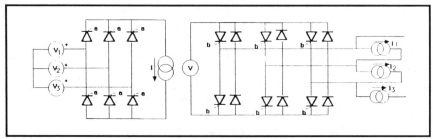

Fig. 2.11 Three-phase all-thyristor Graetz rectifier and its dual circuit

In a voltage rectifier, the control parameter for the voltage across the DC current source is the time the voltage across the thyristors is positive and this voltage can be written:

$$U = \frac{6}{\pi} \sin \frac{\pi}{6} \sqrt{3} \, V_{max} \cos \theta_a \tag{2.8}$$

with θ_a being the turn-on delay angle and V_{max} the amplitude of the voltages V_1, V_2 and V_3.

By duality, in a current rectifier, the current in the DC voltage source is controlled by adjusting the time the current is positive in the dual-thyristor (that is the conduction time of the controlled switch) at the end of which turn-off control is achieved. This DC current $I^*_?$ is:

$$I^* = \frac{6}{\pi} \sin \frac{\pi}{6} \sqrt{3} \, I_{max} \cos \theta_b \tag{2.9}$$

with θ_b being the turn-off delay angle and I_{max} the amplitude of the currents I_1, I_2 and I_3.

This current rectifier has, in fact, the well known structure of a three-phase voltage inverter. Its operation as a line-commutated inverter is, however, less well known since it requires a three-phase AC current source to allow power control by phase-angle regulation.

2.7 CONCLUSION

In this chapter, we have shown that the rules of duality can be applied to static converters and introduced the concept of a dual switch.

Complete application of the duality rules to static converters requires a 'new' switch that bears the dual characteristics to those of a thyristor and will be referred to as a 'dual-thyristor'.

The concept of duality applied to converters is not new and it has already been widely used (with no recourse to the dual switches) in presenting the medium frequency converters such as the thyristor parallel and series inverters (Fig. 2.8a and Fig. 2.9a) [36].

These converters have dual structures but the same commutation mechanisms. In other words, the switches of these converters have dual static characteristics but dynamic characteristics which are followed in the same direction.

On the other hand, in a given structure, two types of switches (that necessarily bear the same static characteristics) with dual dynamic characteristics can be used. Then, as in the case of the thyristor series inverter (Fig. 2.9a) and of the dual-thyristor series inverter (Fig.2.8b), the dynamic characteristics of the switches are followed in opposite directions.

Applied to either the structure or the commutation mechanisms, duality only gives, in most cases, the qualitative properties of these new circuits. By contrast, when applied to both circuits and commutation mechanisms, it also enables transposing the quantitative results of the original converter to the dual converter. Then duality becomes an efficient method of establishing new circuits in power conversion.

So duality should not be taken as a discussion of purely academic interest, but as a real method of study and synthesis for power electronics.

3
RESONANT INVERTERS

3.1 INTRODUCTION

The study of resonant inverters fits into the larger framework of the analysis of providing an AC supply to an electrical load (series RLE) from a DC current or voltage source. The conversion of the electrical energy is achieved by a non-modulated single-phase DC/AC converter whose structure depends on the nature of the DC source. This converter is either a voltage source inverter or a current source inverter that delivers voltage or current 'square' waves.

By virtue of the principle of duality, this presentation of resonant inverters will focus on voltage source inverters without losing generality. After a brief reminder of the principles and the characteristics of operation of these inverters, we will concentrate on outlining their essential features.

3.2 RESONANT INVERTER STRUCTURES

3.2.1 Voltage source supply

The RL load can be directly supplied with voltage by means of a voltage source inverter (Fig. 3.1) composed of four switches which are bidirectional for current and unidirectional for voltage. These switches are made up of a controlled component connected to an antiparallel diode. This controlled component can be voltage asymmetric so that the choice is wide (thyristor, GTO, bipolar transistor or MOSFET for example). Choosing the component requires determining the commutation mechanisms of the switches.

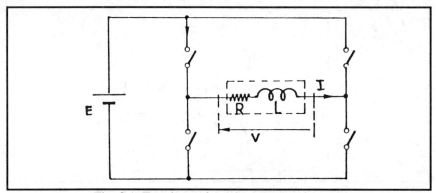

Fig. 3.1 Topology of a voltage-source inverter

The waveforms corresponding to the circuit are shown in Fig. 3.2. The load current is composed of a succession of responses of an RL network to voltage steps. The commutation mechanisms can be derived directly from these waveforms. Each voltage step is caused by turning off a pair of controlled devices. To obtain continuous conduction the controlled devices must be turned on inherently when their antiparallel diode turns off, i.e. as the load current passes through zero.

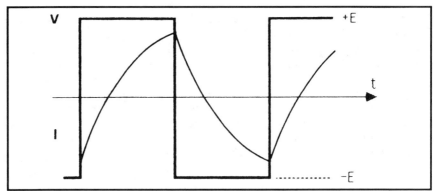

Fig. 3.2 Current and voltage waveforms in a voltage-source inverter supplying a series RL load

In the circuit of Fig. 3.1, dimensioning the semiconductor devices for a given power depends on the 'cos Φ' of the load. In the sine-wave mode, power factor compensation is usually achieved by connecting a capacitor to the load. In the case of supply by a voltage source inverter, this capacitor can only be series connected to the load (Fig. 3.3). Thus the load network is a resonant or oscillatory circuit and the inverter supplying this particular load is known as a **series resonant inverter** or, for short, a series inverter.

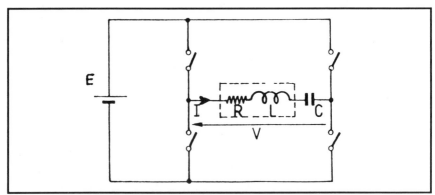

Fig. 3.3 Topology of the series-resonant inverter

In all the presentation of the resonant inverters, z is the damping factor, ω_0 is the angular frequency of the undamped network (or undamped natural angular frequency), and ω is the damped natural angular frequency of the RLC network. In the case of the series network these quantities are:

$$z = \frac{R}{2}\sqrt{\frac{C}{L}} \tag{3.1}$$

$$\omega_0 = \frac{1}{\sqrt{LC}} \tag{3.2}$$

$$\omega = \omega_0\sqrt{1-z^2} \tag{3.3}$$

The load current is now composed of a sequence of responses of the series resonant RLC network to voltage steps. For a given RL load, the nature of these responses obviously depends on the capacitor, i.e. on the damping factor z, and on the switching frequency of the inverter f_s. The different wave forms likely to be observed are shown in Fig. 3.4:

(a) The RLC network is highly damped (z>1). The current and voltage wave forms are similar to those in Fig. 3.2 (infinite C).

(b) The network is oscillatory (z<1) and switching frequency f_s is greater than f, the damped frequency of the network.

(c) The circuit is oscillatory and switching frequency f_s is equal to f, the damped frequency of the network.

(d) The circuit is oscillatory and switching frequency f_s is between f and f/2.

(e) The circuit is oscillatory and switching frequency f_s is less than f/2. After each diode conduction, the current remains at zero: discontinuous conduction mode.

(f) The circuit is oscillatory, the control enables several zero crossing of the load current and $f/(2n) > f_s > f/(2n+1)$ with n=1,2,...

(g) The circuit is oscillatory, the control enables several zero crossing of the load current and $f/(2n+1) > f_s > f/(2n+2)$ with n = 1,2,...

In the two latter cases, n is the whole number of current reversals in a half period of the inverter.

The analysis of these wave forms requires several remarks:

1 • In cases (a), (b), and (f), each voltage step is caused by controlled-device turn-off. Inherent turn-on of the controlled devices occurs as the load current passes through zero. Under these conditions of operation, the controlled devices must be turn-off controlled but can benefit from inherent turn on at zero voltage. Each complete switch may be a dual-thyristor.

On the other hand, in cases (d) and (g), each voltage transition is caused by controlled-device turn on. Controlled device turn off occurs inherently as the load current passes through zero. Each switch can be a reverse conducting thyristor or a thyristor with antiparallel diode.

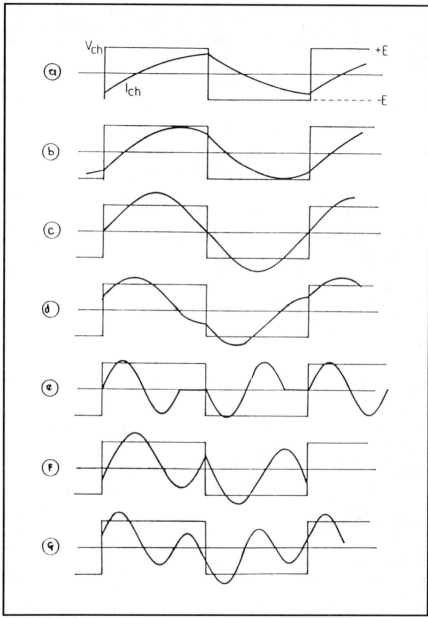

Fig. 3.4 Current and voltage waveforms in a series-resonant inverter for
different parameters of the resonant circuit

Cases (c) and (e) are peculiar operation cases in which the voltage steps occur at the time the load current is zero so that there is no real commutation of the switches. However it should be noted that such operation requires switches to be turn-on and turn-off controlled.

2 • Waveforms (f) and (g) are only possible if the switches have long lasting control signals ('pulse chains'). In any other case the switches require short turn-on or turn-off control pulses. The study carried out in this chapter only deals with this case of short control pulses that necessarily induce operation in the discontinuous mode as soon as $f_s < f/2$.

3 • For a given RL load, the different load current waveforms can be obtained by modifying C and/or the relative values of f_s and f. In other words, it is possible to modify the commutation mechanisms of the inverter, or else to adjust the 'power factor' of the inverter by modifying C and/or f_s.

4 • In the converter structure shown in Fig. 3.3, turn-on controlled switches can only be used if the RLC network is oscillatory. For this reason the resonant inverter provides an application ideally suited to the thyristor, but the load network must have a high quality factor (Q).

The circuits in Fig. 3.5 are variants of the basic 'full bridge' structure of Fig. 3.3 [36].

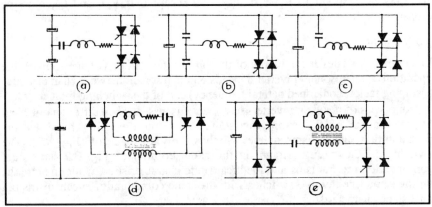

Fig. 3.5 Variants of the series-resonant inverter

3.2.2 Current source supply

In the case of current source supply, the series RL load cannot be supplied directly by a current source inverter; a parallel capacitor is necessary (Fig. 3.6). This capacitor enables the current to be commutated on one hand, and the power factor of the load to be compensated on the other hand.

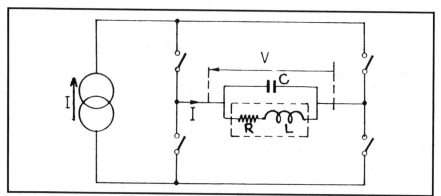

Fig. 3.6 Parallel resonant current-source inverter

The single-phase current-source inverter is made up of four switches, bidirectional for voltage and unidirectional for current. The voltage across the load is then composed of a sequence of responses of a resonant or oscillatory parallel RLC network to current steps. As in the case of the voltage source supply, different waveforms can be obtained depending on the circuit parameters and they correspond to those of Fig. 3.4 except that V_{ch} and I_{ch} must be exchanged.

3.2.3 Conclusion

In summary, whatever the nature of the supply (current or voltage source), the operation of the resonant inverter depends on the resonant network characteristics (damping factor, undamped natural frequency) and of the switching frequency f_s.

Four types of resonant inverters have been discussed. They differ on one hand by the nature of the supplying source, and on the other hand by the commutation mechanisms of the switches they are made of (Fig. 3.7). The switches in the circuits drawn in the same column of Fig. 3.7 have dual commutation mechanisms and identical static characteristics, while the circuits on the same line involve switches with the same commutation mechanisms but dual static characteristics.

Up to now, only the problem of the AC supply of a series RLC load has been discussed. However, the same approach can be applied to an RC parallel load, by adding — instead of a parallel capacitor as formerly — an inductor, which leads to the circuits of Fig. 3.8.

The resonant inverter circuits of Fig. 3.7 and 3.8 are exactly two-by-two dual so that the following presentation only deals with the voltage source resonant inverters.

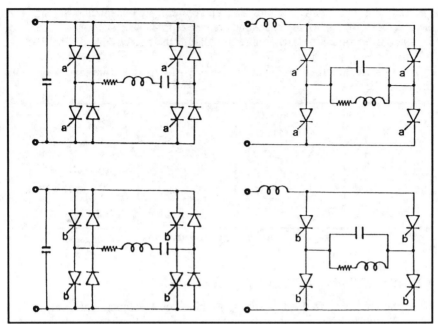

Fig. 3.7 Various structures of resonant inverters with series RL load

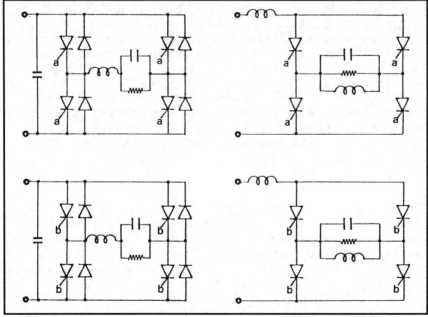

Fig. 3.8 Various structures of resonant inverters with parallel RC load

Assuming that the switches are ideal and commutate instantaneously, and that the passive components such as inductors, capacitors and resistors are ideal also, the study of resonant inverters reduces to that of the two circuits of Fig. 3.9:
(a) - the series resonant network,
(b) - the 'series-parallel' resonant network fed by a square voltage wave with amplitude E.

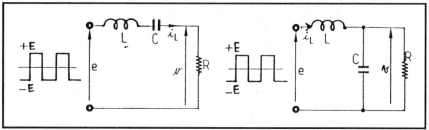

Fig. 3.9 Modelling of resonant inverters

3.3 REVIEW OF RESONANT CIRCUITS

When the switching frequency is close to the natural frequency, the voltage and current wave forms of these circuits are quasi-sinusoidal. Based on the assumption that they are perfectly sinusoidal, which is valid for high Q-factors and in the vicinity of the natural frequency, only the voltage and current fundamentals are involved in the power flow. In this section, we intend to recall briefly a few properties of these circuits in the sine-wave mode.

A particular point of interest is the conditions under which power is transferred in these resonant inverters and thus the V(I) characteristics of the voltage across the load resistance as a function of the current through it. Also, the importance of the commutation processes in the switches within resonant inverters requires the restatement of certain fundamental results concerning the development of the phase relationship between the voltage e and the current drawn from this same source.

The amplitude of the instantaneous quantity x is denoted by X, and its complex amplitude is denoted by $\overline{X}$.

3.3.1 Series resonant network

The frequency response of a series RLC network is shown qualitatively in Fig. 3.10 (it is the resistor current that is illustrated here). The current reaches its peak at the resonant frequency. Frequency variation causes amplitude and phase variations in the current, which produce a consequent variation in the power dissipated in the resistance R. This power control by frequency variation is one of the distinctive features of resonant inverters [37].

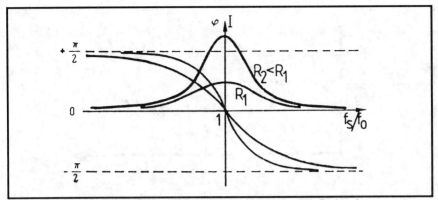

Fig. 3.10 Frequency response of a series RLC circuit

From the vectorial diagrams of Fig. 3.11, the expression of the V(I) characteristic at fixed frequency is readily deduced.

$$V^2 + (I \sqrt{L/C})^2 (u - 1/u)^2 = E^2 \qquad (3.4)$$

where:

$$u = f_s/ f_0 \qquad (3.5)$$

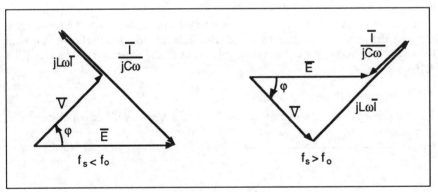

Fig. 3.11 Vector diagrams for the series RLC circuit

This characteristic, which represents the effect of varying the resistance at a given frequency, is an ellipse (Fig. 3.12).

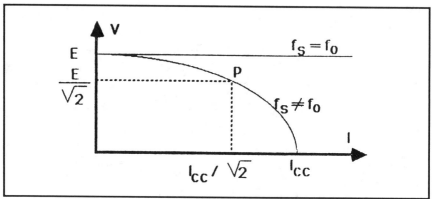

Fig. 3.12 V(I) characteristic of a series RLC circuit

The power delivered by this converter is naturally limited and, for a given frequency, the maximum power able to be transferred to the load is:

$$P_{max} = \frac{E^2\sqrt{C/L}}{4(u-1/u)}$$

(3.6)

This point of operation at maximum power corresponds to **P** in Fig. 3.12. It should be noted that the power factor of the generator supplying this R-L-C network is V/E and equals $1/\sqrt{2}$ at point **P**.

The sign of the phase shift between voltage and current in the inverter that supplies the series resonant network is fully determined by the relative values of the switching frequency of the inverter (f_s) and of the natural frequency f_0 (Fig. 3.10).

3.3.2 Series-parallel resonant network

The frequency response of this circuit is in fact very well known since it is a filter.

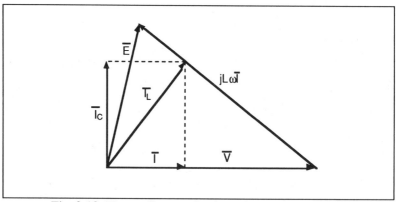

Fig. 3.13 Vector diagram for the series-parallel circuit

The vector diagram related to this circuit is shown on Fig. 3.13 and the expression for the V(I) characteristic is:

$$V^2(1-u^2)^2 + (I\sqrt{L/C})^2 u^2 = E^2 \qquad (3.7)$$

Except for the special cases where the frequency is zero (u=0) (when the voltage across the resistance is constant), and where the frequency is equal to the natural frequency of the undamped network (u=1) (when the current is constant in this resistance), the fixed frequency characteristics are also ellipses (Fig. 3.14).

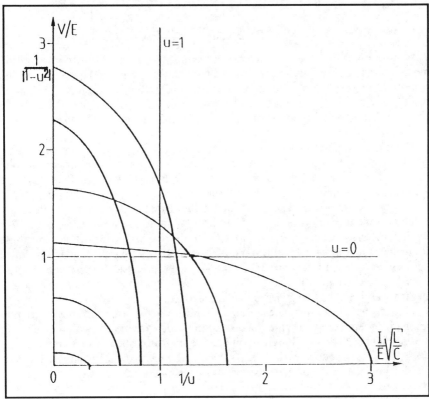

Fig. 3.14 V(I) characteristic for the series-parallel circuit

In contrast to the series resonant network, with a series-parallel resonant network, the sign of the phase shift between voltage e and inductor current I_L depends on the relative values of f_s and f_0 **AND** on the load resistance. An elementary analytic calculation shows that the current delivered by the inverter is always lagging the voltage when u is greater than 1. On the other hand, when u is less than 1, there is a particular value of the load resistance R_n given by:

$$R_n = \sqrt{L/C} \, \frac{1}{\sqrt{1-u^2}}$$

(3.8)

on one side of which the inverter current lags the voltage and on the other side of which the current leads the voltage. When the load resistance is equal to the particular value R_n, the current I has a constant amplitude I_n given by:

$$I_n = E\sqrt{C/L}$$

(3.9)

This value is independent of the frequency of operation.

3.4 PROPERTIES OF RESONANT INVERTERS

Resonant inverters are converters involving switches whose commutation mechanisms are strictly defined. Each of these switches has a controlled commutation, i.e. caused by the control circuit, and an inherent commutation resulting from the influence of the oscillatory circuit on the switch. The consequences of this type of operation are numerous and are reviewed in the following.

The inherent commutation, either turn on at zero voltage or turn off at zero current, theoretically occurs without any loss. As far as the controlled commutation is concerned, the switches can be equipped with snubbers to reduce the stresses applied to them. When the switch is turn-on controlled (e.g. a thyristor), the snubber is a series inductor. Since this switch turns off inherently at the current zero crossing, there is no energy stored in this inductor at turn off (which means no energy to dissipate). Similarly, when a switch is turn-off controlled (dual-thyristor, for instance), the snubber is a parallel capacitor. Since this device turns on inherently when the voltage crosses zero, there is no energy stored in the capacitor at this time. So these snubbers are non-dissipative elements and can thus be generously dimensioned.

Under such circumstances, the switching losses are minimized so that resonant inverters are suitable for operation at high frequency without affecting the efficiency. (The term 'high frequency' refers to frequencies greater than the usual electrical frequencies of 50 and 400Hz). The high frequency chosen depends on the application, on the type of switches used and on the power to be controlled.

This type of operation leads to a reduction in the weight, size and dimensions of the reactive elements (filters, resonant circuit) and of the transformers. Finally, it accelerates the response to transients and makes operation inaudible when the frequency is greater than 20 kHz.

Owing to the favourable commutation conditions of the switches, the stress on the components is reduced to a minimum, which certainly tends to increase reliability. Furthermore, for switching frequencies in the vicinity of the natural frequency, the voltage and current wave forms are sinusoidal and the electromagnetic and radio noise generated by these resonant inverters is reduced. It should be noted that, the smaller the heat to be extracted, the easier it is to provide a screened enclosure.

Two large groups of resonant inverters have been presented:
1 • When the capacitor is series connected to the load the RLC network behaves as an instantaneous current source. The resonant inverters that supply this type of load are voltage source inverters and will be referred to as **series inverters**.
2 • When the capacitor is connected in parallel with the load, the RLC network behaves as an instantaneous voltage source. It must therefore be supplied with current and the resonant inverter used is thus a current source inverter, usually called a **parallel inverter**.

The parallel inverter, which requires components with symmetric voltage capability, is traditionally an application for thyristors (turn-on control) which naturally withstand symmetric voltages and now possibly for GTOs (turn-off control). Because of the intrinsic properties of these components, the parallel inverter is limited to frequencies of a few kHz and is suitable for high power applications (a few MW). For lower powers (a few tens of kW), increasing the switching frequency can be considered provided the function of the thyristor is implemented with fast components, such as MOS or bipolar transistors for instance, in series with a diode. Furthermore, supplying a parallel inverter from a current source makes it remarkably reliable.

The series inverter requires switches with asymmetrical voltage capabilities and the choice is presently wider. Analysis of the commutation phenomena in a voltage source inverter leg shows that the turn-on controlled commutation is more constraining than the turn-off controlled commutation especially because of the technological imperfections of the components (recovery currents in diodes, dV/dt, etc.). If turn-on control has been more widely used and studied, it is because the thyristor has been the only controlled power semiconductor for a long time. Nowadays, with turn-off controlled devices available, a lot of designers still attempt to reconstruct the 'thyristor function' and exclude a priori the possibility of turning them off (not least by including a series inductor limiting the di/dt). This goes against the basic principle of safe operation in a voltage source inverter that involves turning off all the switches in case of a fault. If it is required, nevertheless, to retain the turn-off ability, more complex snubbers must be used. In fact, in a series inverter, it would be best to use switches with controlled turn off and inherent turn on, such as 'dual-thyristors' which among other things, provide a solution to the problem of diode turn off.

Compared to the traditional technique used in pulse width modulated inverters, that involves controlling individually the turn on and turn off of each controlled switch, replacing one controlled commutation by an inherent commutation gives resonant inverters all the above-mentioned advantages. Nevertheless, this modification requires control strategies specific to this type of converter, such as frequency control.

On the other hand, resonant inverters suffer from a great interdependence of the converter and the load. Thus, the study and implementation of these inverters is complex, but they are already widely described in the literature [36,38-42].

Series inverters and parallel inverters are suitable for high frequency operation since their reactive elements are then of low values. At lower frequencies they require large inductor and capacitor values. However, in all applications where the load is highly reactive and when compensation is worthwhile, resonant inverters can be an attractive alternative. The main applications of resonant inverters are 50/400 Hz sine-wave generation, induction heating and ozone production.

3.5 CONCLUSION

The study of the AC supply of a series RL or parallel RC load has demonstrated the main resonant inverter structures.

These converters have three major characteristics that simultaneously constitute their advantages and their disadvantages:

• They deliver unmodulated current and voltage waveforms (simple control of the inverter).

• The addition of a passive element to the load to compensate the 'power factor' also enables a choice of the commutation mechanism within the inverter (increase in price, volume and weight of the equipment but better utilization of the components).

• The inverter can only be controlled by frequency regulation. Independent control of the power and the frequency requires an auxiliary control method (control of the amplitude of the DC source, for example).

Finally, resonant inverters are distinguishable by the 'soft' commutation principle on which their operation is based. Consequently, they have remarkable properties and there is considerable incentive to apply the same techniques to other electrical energy conversion processes.

4

SOFT COMMUTATION

4.1 INTRODUCTION

In the Chapter 3, it has been shown that AC supply of inductive or capacitive loads can be achieved very satisfactorily (high efficiency, low distortion). It has also been shown that the performance of these resonant inverters was tightly linked to the commutation process in these switching devices all of which execute one controlled switching and one inherent switching.

We will define as a soft-commutated direct converter a converter only involving switches, each of these switches having at most one controlled switching and at least one inherent switching, the controlled and inherent switchings being if necessary different from one switch to another. So the soft commutation is not compatible with totally controlled switches. In soft-commutated converters the power transfer is controlled by means of switches or by a set of switches with thyristor or dual thyristor switching features.

The term 'soft commutation' is legitimate since those switches that have one controlled switching and one inherent switching can be fitted with lossless snubbers (a series inductor when turn on is controlled or a parallel capacitor when turn off is controlled) so that the switching losses are considerably reduced.

In this chapter, by considering an elementary commutation cell used in all direct converter configurations, general criteria can be established allowing systematic identification of direct converters able to operate in the soft-commutated mode.

4.2 GENERAL REMARKS ABOUT COMMUTATION IN STATIC CONVERTERS

4.2.1 The elementary commutation cell

The operation of a static converter can be split into a sequence of elementary modes. Each elementary mode is characterised by an electrical network different to the preceding one and derived by modifications to the interconnection of the active branches.

This process of sequential modification of the network calls for a static device composed of star-connected electronic switches (Fig. 4.1). The network branches connected to this static device must obey the rules for the connection of the sources. Thus:

• Each switch is connected to a voltage source (otherwise opening a switch would result in open-circuiting a current source).

45

• The node at the centre of the star is connected to a current source since a voltage source can be connected only to a current source by means of a controlled switch.

• At a given time one and only one switch must be ON to avoid connecting two voltage sources together and open-circuiting the current sources.

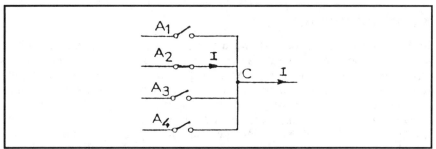

Fig. 4.1 Four way static commutator

The operation of such a device involves a set of complicated phenomena related to the simultaneous and complementary changes of state of two switches; these phenomena are referred to by the general term **commutation**. Consequently, the number of switches composing the static device does not alter the commutation mechanisms as each commutation only involves two switches.

The preceding remarks lead us to consider an elementary cell like that of Fig. 4.2. Voltage source V represents V_1-V_2, the difference of the voltages applied at nodes A_1 and A_2 by the corresponding voltage branches. Current source I represents the mesh current flowing in the branch connected to point C, common to switches K_1 and K_2 involved in the commutation.

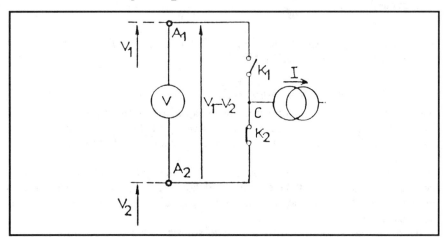

Fig. 4.2 The elementary commutation cell

As an example, Fig. 4.3 shows a few converters with the switches regrouped in order to display the commutation cells.

In an elementary commutation cell, whether the voltage source is voltage-reversible and whether the current source is current-reversible determines the static characteristics of the switches. Whether the voltage sources (respectively current sources) need to be current-reversible (respectively voltage-reversible) or not, depends on the configuration of the converter containing the elementary cell being considered. Hence, the two switches constituting an elementary cell need to have static characteristics featuring the same number of current segments (respectively voltage segments).

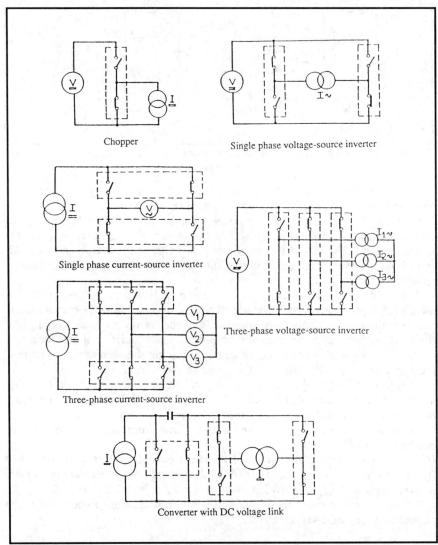

Chopper

Single phase voltage-source inverter

Single phase current-source inverter

Three-phase voltage-source inverter

Three-phase current-source inverter

Converter with DC voltage link

Fig. 4.3 Examples of converters showing the elementary commutation cells

As far as dynamic characteristics are concerned, the commutation of an elementary cell is caused by the controlled switching of one of the switches, which induces the inherent switching of the other. Two commutation modes should be distinguished (Fig. 4.4):
• commutation by turn-on control of the OFF switch (Fig. 4.4a),
• commutation by controlling the turn-off of the ON switch (Fig. 4.4b).

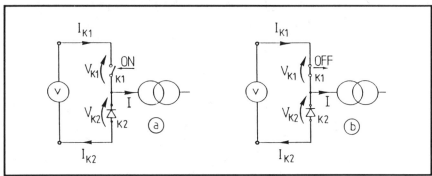

Fig. 4.4 The two possible modes of commutation in a cell: (a) turn-on control
(b) turn-off control

Limiting the analysis to a single switch, since they operate in a complementary way, the following rule can be stated:
If the sign of the current flowing through the switch that is ON prior to the commutation and the sign of the voltage across this switch after the commutation are alike, the commutation must be controlled by turning this switch off. If these signs are unlike turn-on control of the OFF switch is required.

It should be noted that this rule has been established in relation to the ON switch taking account of the voltage and current imposed by the external circuit. So, this rule always applies, independently of the phenomena that could occur during the commutation because of the non-ideal characteristics of the sources.

A more synthetic rule can be obtained by linking the commutation mode to the operation of the converter and the resulting waveforms.

When the commutation results in an *increase in potential at point C* (Fig. 4.2), this commutation will be defined as *positive*; when it results in a *decrease in potential at point C* it will be defined as *negative*.

Defining the current as positive when it flows from point C, the previous rule can be stated as follows:
If the commutation and the current have like signs, commutation is controlled by turning on the switch that is OFF; if the commutation and the current are unlike, commutation is controlled by turning off the switch that is ON.

It must be noted that applying this rule requires strict obedience to the sign conventions defined above.

4.2.2 Different commutation cells

In the special case of AC/AC conversion, the switches are voltage and current reversible. Because of technological limitations, these switches are usually implemented by connecting two three-segment switches in series or in parallel. Furthermore, the voltage source and current source generally have different frequencies and the lower frequency can usually be taken as a slow varying DC compared with the higher frequency. These AC/AC converters can be considered as the juxtaposition of two AC/DC converters and will not be considered in the rest of the study. For this reason, in an elementary commutation cell, at least one of the sources is assumed to be a DC source.

By virtue of the previous remarks, it can be deduced immediately that if one of the switches of the cell is both turn-on and turn-off controlled, the other turns inherently on and off (diode). In all other cases, the two switches must have the same control capabilities: they are only turn-on controlled (or turn-off controlled), the other commutation being inherent. **Five** commutation cells can thus be distinguished.

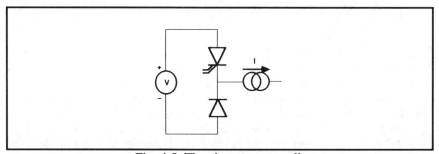

Fig. 4.5 The chopper-type cell

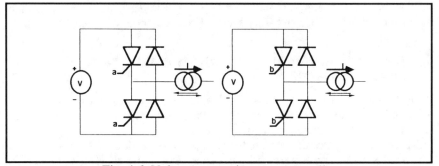

Fig. 4.6 Voltage-source inverter type cells

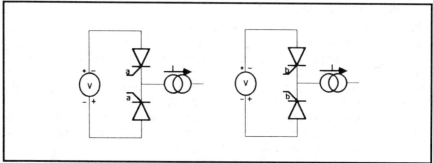

Fig. 4.7 Current-source inverter type cells

• The chopper-type cell (Fig. 4.5) with unidirectional voltage and current sources.
• The voltage-source inverter type cells (Fig. 4.6) in which only the voltage source is unidirectional.
• The current-source inverter type cells (Fig. 4.7) in which only the current source is unidirectional.

4.2.3 Special cases

If the sources are reversible, when either the voltage or the current crosses zero, inherent turn on or turn off of one of the switches can occur, thus causing synchronous inherent switching of the other. This phenomenon is general and occurs each time a controlled switching is replaced with an inherent switching or is suppressed. Under such circumstances, the commutation is performed without any control and only depends on the natural evolution of a given electrical quantity. Such a commutation will be referred to as a free commutation; it occurs in diode or thyristor-diode bridges for example.

In some cases, one switch may operate by itself in a circuit branch. But under such circumstances, one can hardly speak of commutation (dimmers).

4.3 SOFT COMMUTATION IN AN ELEMENTARY COMMUTATION CELL

With the help of the rules stated above and according to the types of sources involved we now have to establish the conditions necessary for soft commutation in an elementary commutation cell.

The elementary commutation cells capable of operation in the soft mode, in which the switches must necessarily have the same static characteristics and the same commutation mechanisms, are shown in Fig. 4.6 and 4.7.

Still using the notation of Fig. 4.2, the commutation from K_2 to K_1 and voltage V are alike while commutation from K_1 to K_2 and voltage V are unlike. The soft commutation mode is then characterized by the following relation:

$$(sgn(V)^*sgn(I))_{K_2 \to K_1} = (sgn(V)^*sgn(I))_{K_1 \to K_2} \qquad (4.1)$$

Obviously, this relation is only fulfilled if either V or I changes sign between two successive commutations. Two conditions required for operation in the soft commutation mode can be inferred at once:

•At least one of the voltage and the current sources must be reversible (the voltage source must be voltage-reversible and/or the current source must be current-reversible).

• The switching frequency of the cell is rigidly constrained and generally equal to the zero-crossing frequency of the quantity that enables soft commutation.

Limiting the study to a single AC source, two main types of commutation cells capable of soft-mode operation can be distinguished:

(a) current commutated cells with soft commutation enabled by the reversal of the current (Fig. 4.6),

(b) voltage commutated cells with soft commutation enabled by the reversal of the voltage (Fig. 4.7).

Furthermore, in each cell of Figs 4.6 and 4.7 the controlled switching of one or both switches can be replaced with an inherent switching or suppressed, which amounts to replacing this or these switches with one or two diodes.

Lastly, it should be noted that all these commutation cells are duals of each other, pair by pair. A current-commutated cell controlled at turn off (respectively at turn on) is the dual of a voltage-commutated cell controlled at turn on (respectively at turn off).

4.4 EXAMPLES OF CONVERTERS OPERATING IN THE SOFT COMMUTATION MODE

4.4.1 DC/DC conversion

Such a converter in which the current source must be current-reversible if the input is a DC voltage source, must be composed only of current-commutated cells to operate in the soft-commutation mode. The topology of this DC/DC converter is that of a current-reversible chopper (Fig. 4.8a) or that of a four-quadrant chopper in which both the current and voltage can reverse (Fig. 4.8b).

Consider the circuit of Fig. 4.8a. Voltage v applied across the LC filter is unidirectional. Power flow occurs only if current i_L has a non-zero DC component. On the other hand, operation in the soft-commutation mode implies that i_L changes sign at each commutation. Thus the AC component of the current must be greater than its DC component.

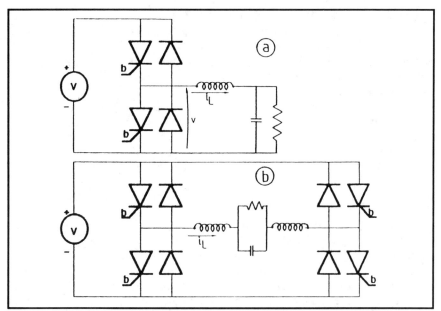

Fig. 4.8 DC/DC converter: (a) current reversible, (b) current and voltage reversible
(Soft commutation is compatible with PWM as soon as the ripple on current i_L is greater than the average value of this current over a switching period.)

In the chopper of Fig. 4.8a, these conditions can be obtained, for example, by using a hysteresis current control. However, one must note that only the turn-off control enables effective control of the current amplitude [43].

This principle of operation can obviously be extended to the four-quadrant chopper of Fig. 4.8b, which finally leads to pulse-width-modulated DC/AC conversion.

4.4.2 DC/AC conversion

Continuing with the case of supply from a DC voltage source, the most general topology allowing DC/single phase AC conversion is the bridge composed of two elementary current-commutated cells. Each of these cells commutates in each half-period of the current. Two principal soft commutation modes of operation of this single phase inverter are to be considered (Fig. 4.9):

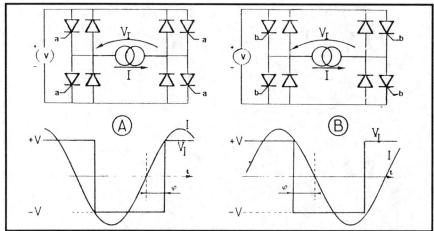

Fig. 4.9 The two modes of soft commutation operation in a single-phase
inverter

• All the commutations are turn-on controlled (Fig. 4.9a). The fundamental of voltage V_I lags current I.

• All the commutations are turn-off controlled (Fig. 4.9b). The fundamental of voltage V_I leads current I.

The converters of Fig. 4.9 share the following properties:

• Power is regulated by controlling the phase shift φ between I and the fundamental of V_I.

• The soft commutation mode implies that power can be transferred in both directions and that DC voltage source E is current-reversible.

• Extension to DC/polyphase AC converters is possible.

In each of the circuits of Fig. 4.9, the controlled switching of one switch of each cell can be suppressed, or else two switches of the same cell can be replaced with diodes (Fig. 4.10). Both these solutions yield the same waveforms.

The converters of Fig. 4.10 share the following properties:

• The power flow is regulated by controlling the width of the zero-volt plateau in the waveform of voltage V_I.

• The power flow is unidirectional and can only be from the AC source to the DC source (rectifier operation). The DC voltage source can be unidirectional for current.

• Extension to multiple phase converters is possible to the extent that, for reasons of symmetry, each cell consists of one controlled switch and one diode.

• In the particular case when φ=0 we have the diode bridge.

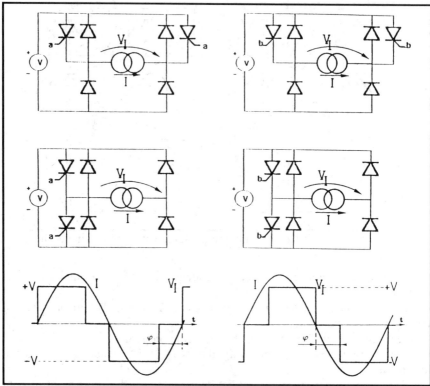

Fig. 4.10 Various forms of half-controlled (or mixed bridge) non-reversible
current rectifiers

A final type of converter is obtained by using two cells with dual
commutation mechanisms (Fig. 4.11). The commutations are shifted and the
current necessarily crosses zero during the zero-volt plateau of voltage V_I
[20,22,24]. The power can flow from the DC source to the AC source (inverter
mode) but the rectifier mode can also be achieved. Nevertheless, whatever the
operation mode, the power flow is unidirectional and controlled by adjusting φ,
the width of the zero-volt plateau in voltage V_I. It would be difficult for this
type of operation to be extended to polyphase converters (except for the special
case of totally unbalanced loads).

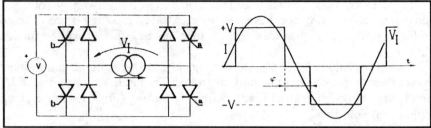

Fig. 4.11 Single-phase inverter with legs involving switching devices with dual
dynamic characteristics

4.4.3 Direct DC/AC/DC conversion

The block diagram of such a converter is shown in Fig. 4.12. The input is a DC
voltage source while the output is a DC current source (in order to comply with
the basic rules of static converters) [20,31,44].

Inverter CS_1 delivers an AC voltage denoted V_a that allows soft
commutation of rectifier CS_2 while CS_2 can, with respect to CS_1, be taken as
an AC current source I_a that allows soft commutation of CS1. The switching
frequency of CS_1 and CS_2 must then be equal.

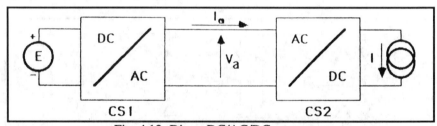

Fig. 4.12 Direct DC/AC/DC converter

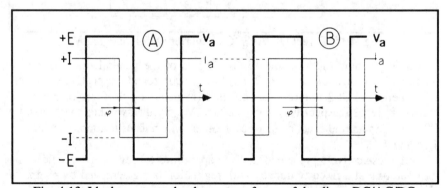

Fig. 4.13 Ideal current and voltage waveforms of the direct DC/AC/DC
converter

To the extent that power is intended to flow from the DC voltage source to the current source, CS_1 operates as an inverter and its topology is one of those in Fig. 4.9. The case when CS_1 uses the topology of Fig. 4.11 will be studied further on.

Correct operation of the converters in Fig. 4.9 strictly requires a bidirectional power flow so that CS_2 must be fully controlled. The idealized waveforms of V_a and I_a are shown on Fig. 4.13 with current I_a leading (Fig.4.13a) or lagging (Fig.4.13b) voltage V_a.

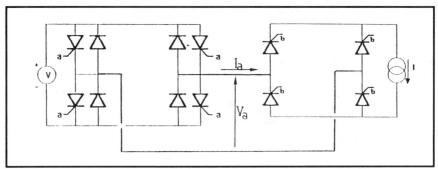

Fig. 4.14 Direct DC/AC/DC converter with I_a leading V_a

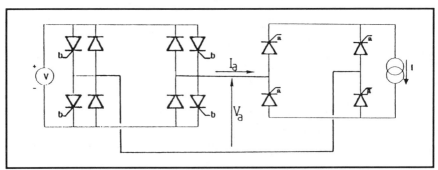

Fig. 4.15 Direct DC/AC/DC converter with I_a lagging V_a

Taking account of the conventions chosen, of the general rules set out in section 4.2.1 or else of Fig. 4.9, CS_1 is turn-on controlled (respectively turn-off controlled) and CS2 is turn-off controlled (respectively turn-on controlled) if current I_a leads (respectively lags) voltage V_a. So the switches that make up CS_1 and CS_2 are the exact duals of each other, which leads to topologies of Fig. 4.14 and 4.15.

Both these converters have attractive capabilities for reverse power flow and for operating at a fixed frequency with regulation being achieved by means of phase shifting (Fig. 4.13).

If only unidirectional power flow is required, one is naturally inclined to use a half-controlled, or even an uncontrolled (diode) rectifier for CS_2. Then the only inverter that could possibly operate under such circumstances is that of Fig. 4.11. In practice, however, because of the voltage drops in the switches, operation cannot be achieved unless the switches in CS_2 are turn-on **AND** turn-off controlled,since:

• if the CS_2 switches turn on inherently, then the dual thyristors in CS_1 cannot commutate [45-48],

• if the CS_2 switches are turn-on controlled (thyristors), the e.m.f. required to commutate these thyristors is not present.

Nevertheless, it should be noted that all commutations of the CS_2 converter occur under zero (or near zero) voltage conditions.

4.5 SOFT COMMUTATION AND POWER CONTROL

Taking account of the soft commutation criteria and of the few examples quoted above, we can now study the consequences of the soft commutation mode of operation on methods of power flow control.

The techniques commonly used to regulate the power flow are:

1 • pulse-width-modulation (PWM) in DC/DC and DC/AC converters, and more seldom in AC/AC converters,

2 • phase shift control in AC source-commutated inverters and rectifiers as well as in AC/AC converters (cycloconverters).

The term self-commutated ('autonomous') inverter refers to DC/AC inverters that dictate the frequency and possibly the amplitude of the AC. On the contrary an AC-source-commutated ('non-autonomous') inverter is a converter with the switching frequency imposed by the AC source; the AC source can also impose the amplitude of either the current or the voltage. The power flow orientation determines the inverter or rectifier designation.

Pulse-width-modulation is entirely compatible with soft commutation in circuits such as those of Fig. 4.8, provided the current ripple is greater than the average value of this current. However this operating mode suffers from several drawbacks:

1 • The high ripple of the current requires the switches to be highly overrated (by a factor greater than two) as far as instantaneous current is concerned.

2 • In the case of DC/AC conversion (Fig. 4.8b) the frequency of the output must be limited to keep the commutation capabilities of the switches. Furthermore, if a transformer is required, it cannot benefit from high frequency operation since its design depends on the frequency of the output.

Reducing the current ripple makes the soft commutation mode impossible in DC/DC converters and incompatible with pulse-width-modulation in DC/AC converters.

Assuming always that the AC source is sinusoidal, the only waveforms that could be obtained in a soft commutated DC/AC converter are those shown in Fig. 4.9, 4.10 and 4.11 and their duals.

When the converters are operating as rectifiers or as non-autonomous inverters, power control is achieved naturally by phase-shifting. Notice that the circuit in Fig.4.11 must be considered as a non-autonomous inverter as the operating frequency is fixed by the current source if the required commutation conditions are to obtain. (If the current source is, in fact, an RLC circuit, then the frequency is equal to the damped natural frequency of that circuit.)

In self-commutating inverters (Fig. 4.9), the phase shift between the voltage and the current of the AC source is governed by this source. Thus the control only acts on one parameter: the frequency.

Frequency variation can be a versatile way to control power due to the consequent impedance variation. However, it is often incompatible with the practical uses of DC/AC converter (motor drives, induction heating, etc.). Thus, recourse is made to DC source amplitude variation or to more complicated processes such as those discussed in Chapters 5 and 6.

4.6 CONCLUSION

In this chapter, the basic laws governing commutation in converters have been stated. Applying these laws with respect to the operating conditions enables the determination of the commutation mode of each elementary cell of a converter.

These rules have also allowed the compilation of a list of the varied topologies capable of soft commutation operation.

The main condition a cell should satisfy in order to operate in the soft commutation mode is to be connected to at least one reversible source. Furthermore, the operating frequency must be tied firmly to the frequency of reversal of this source.

Several examples of soft switched converters have been listed. In each of these converters the commutations can be either turn-on or turn-off controlled, though the choice of the controlled switching gives the converter specific properties.

In effect, in a current-commutated cell (Fig. 4.6), after a transient, the direction of the current at the switching time might be wrong. Under these circumstances, if the switches are turn-off controlled, the only possibility is to try and turn off a switch while its antiparallel diode is conducting. Control of the converter is lost, but this occurs during a freewheeling or regenerative phase, which limits the consequences for the converter. On the other hand, in a turn-on controlled converter, fatal shorting of the voltage source through the switches of the cell can arise.

Thus, turn-off control gives a natural safety of operation to converters using current commutated cells. On the other hand, 'safe' turn-on control can only be achieved by means of electronic control that only allows a switch to be turned on if the current has already reversed (and for a sufficient time in the case of thyristors) in the complementary switch and if this latter switch can turn off at the current zero-crossing.

Applying duality rules leads to dual conclusions about voltage commutated cells.

Hence, soft commutation is achieved more satisfactorily with turn-off controlled current commutated cells (voltage inverters) and turn-on controlled voltage commutated cells (current commutator).

These conclusions are totally compatible with the practical behaviour of the sources, since:
• AC voltage sources, except for parallel resonant networks, usually include a series inductor that acts as a turn-on snubber (commutation inductor).
• DC voltage sources, if they have a series inductance, are more easily decoupled than AC sources.

These conclusions also confirm the analysis carried out on the utilization of the components in resonant inverters.

5

FORCED COMMUTATION

5.1 INTRODUCTION

After establishing the soft commutation conditions and giving a few examples of converters operating in such a mode, we will see in this chapter and in the next one how the different types of energy conversion can be reconsidered using only this soft commutation.

Except for some particular cases, soft commutation can only be considered in the unmodulated DC/AC converters, among which one should distinguish:

• AC source-commutated inverters and rectifiers in which the commutation mechanisms (and consequently their reactive behaviour) can be chosen a priori, and

• self-commutating inverters in which the commutation mechanisms are imposed by the load they are supplying.

So, soft commutation is excluded in three large groups of converters:

1 . DC/DC converters,

2 . DC/AC converters when the commutation mechanisms are incompatible with the load characteristics,

3 . modulated converters that behave as DC/DC converters in some regions of operation.

Two techniques nonetheless enable soft commutation to be used in these converters.

The first method resulting from the application of a logical synthesis procedure, involves equipping the converter with all the auxiliary circuits necessary to guarantee the local conditions that enable the switches to soft commutate, independently of the conditions imposed by the sources to which the converter is connected. This approach is not really new since it was widely used when the thyristor and the diode were the only switching devices available, these switches operating necessarily in the soft switching mode. This is **forced commutation** [49~51].

The second method involves conceiving 'new' conversion processes and methods of power control compatible with soft commutation. To do this, particularly in the case of DC/DC conversion, an intermediate AC stage must be created that enables soft commutation conditions in each of the converters it is connected to. An example of such a converter was given in the previous chapter. This second approach allows the whole converter to operate in soft commutation conditions, and more generally, it leads to systems of **conversion with an intermediate AC stage** [52,53] which will be studied in the next chapter.

5.2 FORCED COMMUTATION IN DC/DC CONVERTERS

The basic structure of a DC/DC converter is shown on Fig. 5.1 (step-down chopper). Switch K_p is a two-segment switch with unidirectional voltage and current capabilities that must also be turn-on and turn-off controlled.

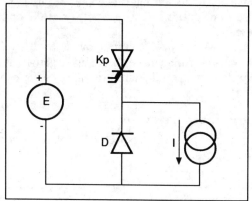

Fig. 5.1 Step-down chopper

To operate in the soft commutation mode, K_p must be chosen from among the four switches able to operate in such a mode. These switches, of which the static characteristics all include the two segments of switch K_p, are reviewed in Fig. 5.2.

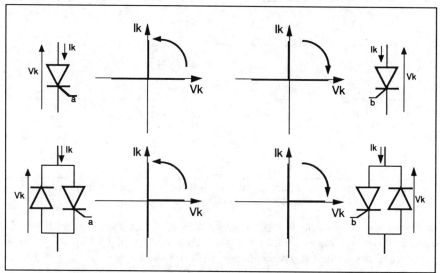

Fig. 5.2 Switches capable of operating in the soft commutation mode

61

However, using one of these switches intrinsically introduces a commutation problem, because of its inherent commutation:
• at diode turn on, if K_p is turn-on controlled (K_p->D),
• at diode turn off, if K_p is turn-off controlled (D->K_p).
Furthermore, two very distinct cases should be considered:
• If K_p is bidirectional with respect to voltage, the inherent commutation only occurs if the voltage across its terminals becomes negative, at the beginning of the blocking phase to enable inherent turn off, or at the end of this phase to enable inherent turn on.
• If K_p is bidirectional with respect to current (apply duality rules!), the inherent commutation only occurs if the current becomes negative, at the beginning of the conduction phase to enable inherent turn on, or at the end of this phase to enable inherent turn off.

5.2.1 K_p bidirectional with respect to voltage

Consider the circuits in Fig. 5.3 that respectively involve turn-on controlled and turn-off controlled switches.

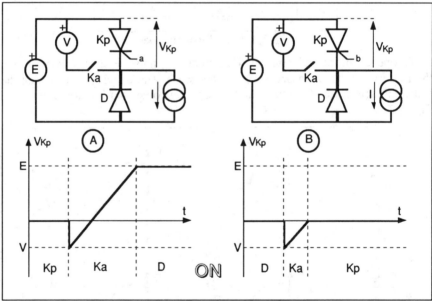

Fig. 5.3 Forced commutation of a switch K_p with bidirectional voltage capability

Following from what has been stated above, commutation is obtained by connecting an auxiliary negative voltage source V in parallel with K_p by means of an auxiliary switch K_a when turn off (Fig. 5.3A) or turn on (Fig. 5.3B) is desired.

The switch K_a, whose turn on is necessarily controlled since it causes the inherent turn off of the conducting switch, that is to say K_p (Fig. 5.3A) or D (Fig. 5.3B), should, like K_p, turn off inherently to experience soft commutation. As a consequence, the inherent turn on of the blocking switch, that is to say diode D (Fig. 5.3A) or switch K_p (Fig. 5.3B) can only occur if the voltages of the commutation cells involved in each of these circuits change sign (free commutation). Then voltage V must rise from its initial negative value at the instant K_a turns off, as shown in Fig. 5.3. Incidentally, the voltage source V can readily be recognised as a pre-charged capacitor.

Thus, the process of forced commutation described above involves replacing the original EK_pD cell with two elementary soft commutation cells:
• the one intended to ensure inherent turn off of the conducting switch following the turn on of K_a, and
• the other intended to allow free commutation from Ka to the incoming switch.

Thus, it can be demonstrated that forced commutation is in fact composed of a natural commutation and a free commutation.

These forced commutation circuits are obviously incomplete since, after the commutation, the capacitor has the wrong polarity (Fig. 5.3A), or is fully discharged (Fig. 5.3B). Many circuits have been proposed in the literature [49,50]to solve these problems, mainly in the case of a turn-on controlled switch K_p (thyristor). All of them make use of at least one extra inductor to reverse the capacitor voltage.

In this section we shall review a few standard forced commutation circuits that use at the most one auxiliary switch, as well as some of their properties based on the state plane described in section 6.3.1.1. The most well known of these circuits are those in Fig. 5.4, without doubt.

The circuit in Fig. 5.4A is well known for its poor characteristics, the voltage V_{Co} available across capacitor C to turn off K_p being a decreasing function of I. On the other hand, in the circuit of Fig. 5.4B, in the which K_p is turn-off controlled, the same forced commutation circuit produces a far better performance. In this case, voltage V_{Co} increases as the load current increases and the oscillatory network is supplied with energy at an appropriate time, that is to say at the commutation from D to K_a.

This fundamental self-adapting property [50], of the voltage V_{Co} to the amplitude of the load current can nonetheless be obtained in the circuit of Fig. 5.4A provided the energy stored in the resonant circuit is not allowed to return to source E during the free-wheeling phase. This can be achieved in two different ways [49], :
• by means of a series diode D_s (Fig. 5.5A),
• by exchanging K_a and D_a (Fig. 5.5B).

The operation of the circuits in Fig. 5.5 only differ by the time at which the capacitor voltage inversion begins: at the turn on of K_p (Fig. 5.5A), or at the turn on of K_a prior to K_p turning off (Fig. 5.5B).

In these circuits (Fig. 5.4 and 5.5), controlling the turn on of K_p and K_a enables control of the duration of the free-wheeling phase and of the active phase, which gives independent control of the power and the frequency.

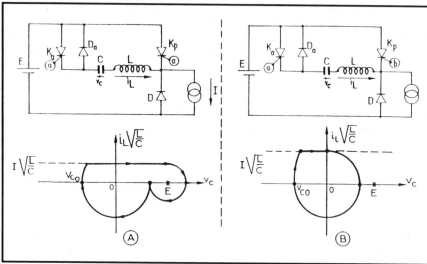

Fig. 5.4 State plane analysis of (a) a turn-on controlled forced commutated chopper, and (b) a turn-off controlled forced commutated chopper

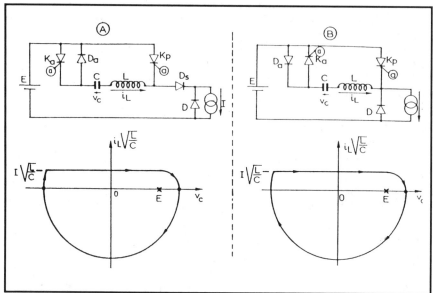

Fig. 5.5 Self-adapting forced commutated chopper

In the interests of simplification, the controlled turn on of switch K_a can also be replaced with an inherent turn on, which amounts to bypassing switches K_a and D_a. However this approach can only be satisfactorily applied to the circuit of Fig. 5.5A. In the resulting circuit which is shown in Fig. 5.6, a degree of freedom is lost and the frequency cannot be controlled independently of the power. The frequency becomes the parameter that controls the power.

Though the circuit in Fig. 5.4B admits of hardly any variation, that of Fig. 5.4A leads to two variants shown in Fig. 5.7. They are derived by moving the inductor to be in series with switch K_p and then by exchanging K_a and D_a.

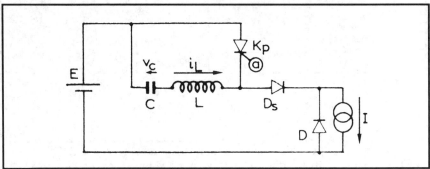

Fig. 5.6 Self-adapting forced commutated chopper without an auxiliary thyristor

Since current I must satisfy the following condition:

$$I < E\sqrt{C/L} \qquad (5.1)$$

to cause the current in switch Kp to go to zero, these circuits (Fig. 5.7) only have a limited commutation capability. In these circuits, inductor L is no longer able to supply an energy proportional to the load current. However, it now performs a new function, that of a snubber for switch K_p (dI/dt limiter or commutation inductor). It should be noted that in the circuits of Figs 5.4, 5.5, and 5.7, inductor L always acts as a snubber for switch K_a.

Replacing switches K_a and D_a with a short circuit in the converters of Fig. 5.7 leads to the circuit of Fig. 5.8 that, like that of Fig. 5.6 and for the same reasons, operates at a variable frequency.

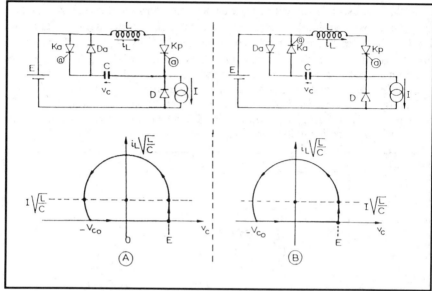

Fig. 5.7 Chopper with limited commutation capability ($I < E\sqrt{C/L}$). The inductor limits the dI/dt in the thyristor and thus the switching losses

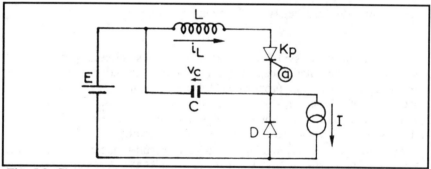

Fig. 5.8 Chopper with limited commutation capability and with no auxiliary thyristor: unidirectional quasiresonant zero current switching chopper.

5.2.2 K_p bidirectional with respect to current

DC/DC converters (step-down choppers) making use of switches with bidirectional current capability and controlled at either turn on or turn off are shown on Fig. 5.9.

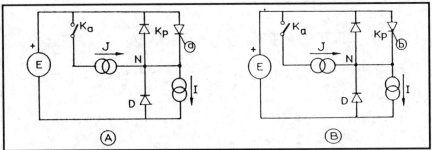

Fig. 5.9 Principle of forced comutation in a chopper using a switch bidirectional with respect to current (turning switch K_a off inhibits current source J)

The forced commutation process used in these converters usually involves an auxiliary current source J that is superimposed with current I via node N in such a way as to cancel inherently the current in the conducting switch, that is to say in K_p (Fig. 5.9A) or in D (Fig. 5.9B).

In practice, this current source J is realised by discharging a capacitor, charged in a previous phase, through an inductor. From which it can be deduced that such a process should logically involve an auxiliary switch K_a that is necessarily turn-on controlled to release the energy stored in the capacitor, and an inductor connected in series in the commutation loop.

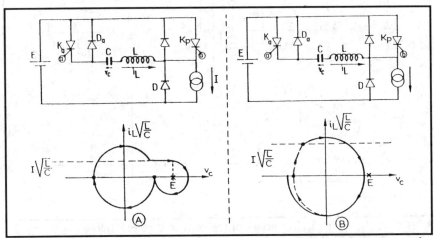

Fig. 5.10 State plane analysis of (a) a turn-on controlled forced commutated chopper, and (b) a turn-off controlled forced commutated chopper

The circuits of Figs 5.10, 5.11 and 5.12 are variants of the circuits in Figs 5.4, 5.5 and 5.7 derived by connecting an antiparallel diode to K_p.

In the circuits of Fig. 5.10B and 5.11 there is no mode with a constant current charging the capacitor and the loci in the state plane are limit cycles whose stability is entirely due to the losses in the LC network and in switches.

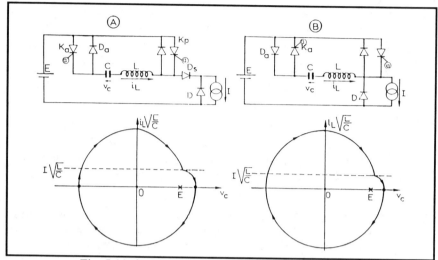

Fig. 5.11 Self-adapting forced commutated chopper
(using for K_p a switch which is bidirectional with respect to current suppresses
the mode in which capacitor C is charged with a constant current)

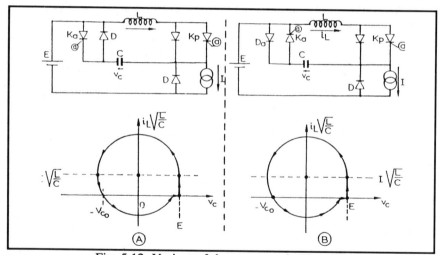

Fig. 5.12 Variant of the converter in Fig. 5.7

On the contrary, in the circuits of Figs 5.10A and 5.12 there is a mode with a constant current charging the capacitor, which in effect constitutes a resetting of the commutation network.

Removing whole or part of the constant current charge modes, results in duration of the commutations being effectively determined by the values L and C and more or less independent of the load current.

At the expense of variable frequency operation, switches K_a and D_a can also be removed from the circuits of Figs. 5.11A and 5.12. The circuits of Figs 5.13 and 5.14 are derived from those of Figs 5.6 and 5.8 respectively, with K_p being bidirectional with respect to current.

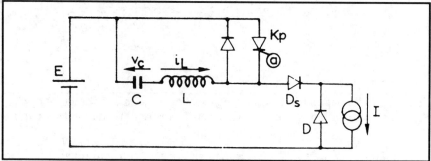

Fig. 5.13 Self-adapting forced commutated chopper with no auxiliary thyristor

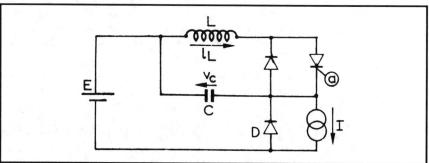

Fig. 5.14 Chopper with limited commutation capability and with no auxiliary
thyristor: quasiresonant bidirectional ZCS chopper

5.2.3 Conclusion

The analysis of the various converters discussed so far in this chapter show that their characteristics are related directly to the placing of the inductor which in all cases is present to cause the reversal of the polarity on the capacitor C when K_p is turn-on controlled:

• Connected in series with capacitor C, it gives the forced commutation circuit the 'self-adapting property' (except for the special case of Fig. 5.4a).
• Connected in series with Kp, it acts as a turn-on snubber for this switch, but the commutation ability is limited.

Investigation of a circuit that combines both these properties leads to a converter (Fig. 5.15) involving two inductors and an auxiliary controlled switch K_a, the operation of which is described in some publications [54—55].

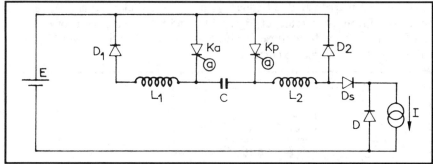

Fig. 5.15 Self-adapting forced commutated chopper with dI/dt limiter for switch K_p (the duration of the mode in which a constant current is charging capacitor C is also minimized)

In addition, two DC/DC converters using a turn-off controlled switch K_p operating in the forced commutation mode have been introduced (Figs 5.4B and Fig. 5.10B). In both these circuits, redrawn in Fig. 5.16, we find the series LC resonant network with its intrinsic properties and the switch K_a. However, replacing a controlled turn on with a controlled turn off requires a second capacitor to assist with the opening of switch K_p. These circuits might be an attractive alternative to the circuits of Fig. 5.15.

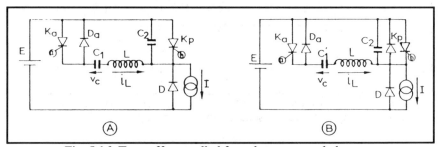

Fig. 5.16 Turn-off controlled forced commutated chopper

5.3 FORCED COMMUTATION AND DUALITY

All the forced commutation circuits presented up to now share two common characteristics:
• The objective is usually the turn off of the conducting switch (the weight of history).
• The energy required to induce the inherent commutation of the controlled switch is stored in a capacitor.

In the circuits where K_p is bidirectional with respect to voltage, the commutation problems have been solved within the commutation cell. K_p is regarded as forming part of two commutation cells that deliver the same current and the variation of the voltage (which is the commutating variable in this case) in each of these cells enables the forced commutation by causing first a natural commutation and then a free commutation.

A different approach was adopted in the circuits with K_p bidirectional with respect to current. Reasoning was directly based on the commutating quantity (the current) in a single commutation cell. The current delivered by this cell is made bidirectional by summing with an auxiliary current source.

To make complete this chapter about forced commutation, we must reconsider a few points:

1 . Could not forced commutation be oriented towards the turn on of the blocking switch?

2 . Is it not possible to store in an inductor the energy required to induce inherent commutation of the controlled switch?

3 . When K_p is bidirectional with respect to current, can we not also investigate the commutation cell?

4 . When K_p is bidirectional with respect to voltage, can we not directly influence the commutating quantity (the voltage) in a single commutation cell?

Obviously duality answers these different questions.

It can already be concluded that storing the commutation energy in an inductor requires an auxiliary switch with turn-off control to release this energy.

Applying duality rules to the circuits presented previously leads to current-fed choppers of the step-up type. To keep this presentation homogeneous, the results obtained have been transposed to step-down choppers. This explains the few differences which might appear between two dual structures.

5.3.1 K_p bidirectional with respect to current

The principle of operation of commutation circuits aimed at turning on the blocking switch, K_p when it is turn-off controlled (Fig. 5.17A) or diode D when K_p is turn-on controlled (Fig. 5.17B), is based on a current source with appropriate sign, series connected in the commutation loop and previously shorted by an auxiliary switch K_a (Fig. 5.17). The commutation is then initiated by opening auxiliary switch K_a which causes a negative current flow in switch K_p when we want to turn it on (Fig. 5.17A) or off (Fig. 5.17B).

Switch Ka is necessarily turn-off controlled since its blocking causes inherent turn on of the blocking switch, i.e. Kp (Fig. 5.17A), or D (Fig. 5.17B); it must then also turn on inherently to operate in the soft commutation mode.

71

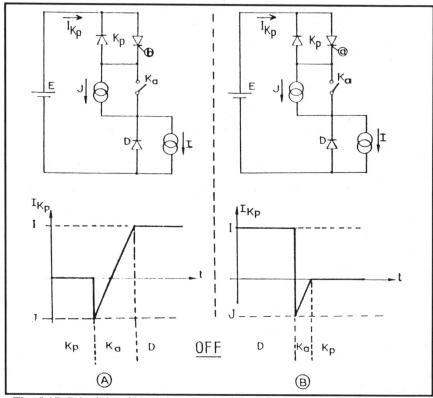

Fig. 5.17 Principle of 'dual' forced commutation with switches bidirectional
with respect to current

Consequently, the inherent turn off of the conducting switch, that is to say diode D (Fig. 5.17A) or switch K_p (Fig. 5.17B), can only occur if the current in the commutation loops involved at that time in each of the circuits changes sign (free commutation). Then current J must rise from its negative value corresponding to the turn off of K_a as shown in Fig. 5.17. In passing, this current source J can readily be identified as an inductor carrying an initial current.

For actual operation, these circuits need elaborating in order to allow reversal of the inductor current. This can be achieved, for example, by replacing K_a with a switch which is voltage reversible and turn off controlled, and by connecting a capacitor in parallel with inductor L or with switch K_p.

Following these principles, it is possible to derive all the dual versions of the circuits of Figs 5.4 to 5.8. Here, study will be limited to those using only one auxiliary switch (Fig. 5.18).

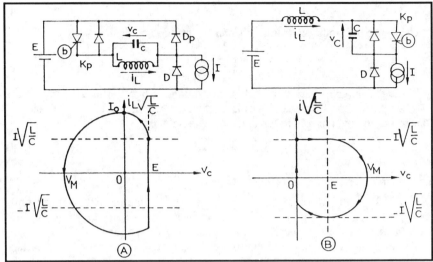

Fig. 5.18 Dual forced commutated chopper with no auxiliary switch: (a) "self-adapting" converter and (b) unidirectional quasiresonant zero voltage switching converter(proper operation requires a minimum current $(I > E \sqrt{C/L}\,)$)

In the circuit of Fig. 5.18A, the current available in the inductor to turn K_p on is:

$$I_0 = I + E\sqrt{C/L} \tag{5.2}$$

So, with this structure, the "self-adapting property" of the forced commutation circuit is recovered.

After the turn on of K_p, the current rises linearly in inductor L with a slope (E/L) that does not depend on the load current. However, this mode which finishes when the free-wheel diode turns off, has a duration that increases with the load current. The behaviour of this circuit is comparable to that of Fig. 5.6 in which the constant current charging of the capacitor had a duration that decreased 'twice' as fast as the load current because it is a function of both the slope and the amplitude.

Proper operation of the circuit in Fig. 5.18B requires the following condition to be satisfied:

$$I > E\sqrt{C/L} \tag{5.3}$$

Thus, there is a minimum value of current below which control of the converter is lost, but during the free-wheeling phase. Capacitor C acts in this circuit as a turn off snubber for switch K_p. This operation is similar to that of the circuit in Fig. 5.8.

These circuits (Fig. 5.18) suffer equally from a major drawback, that is the overvoltage which the capacitor must withstand, which, in each case, is:

$$V_M = E + I\sqrt{L/C} \tag{5.4}$$

Using the previous condition, this can be written as:

$$V_M = E\left(1 + \frac{I}{I_{min}}\right) \tag{5.5}$$

Thus, the range of variation of this overvoltage is tightly linked to that of the load current I. This property is similar to that of the circuits in Fig. 5.6 and 5.8 concerning the overcurrent in the inductor which reads:

$$I_M = I + E\sqrt{C/L} \tag{5.6}$$

5.3.2 K_p bidirectional with respect to voltage

The DC/DC converters of the step-down type implemented with current-reversible switches which are turn-on or turn-off controlled are shown in Fig. 5.19. In these circuits the commutating quantity is the voltage.

The process of forced commutation in these converters requires an auxiliary voltage source series connected with source E so as to ensure the inherent turn on of the blocking switch, that is to say diode D (Fig. 5.19A) or switch K_p (Fig. 5.19B).

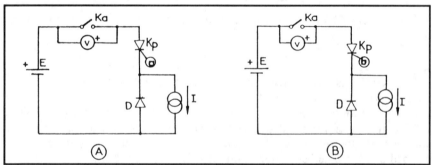

Fig. 5.19 Principle of dual forced commutation with switches bidirectional with respect to voltage

In practice, voltage source V is a capacitor charged by a current previously established in an inductor. This current flow is stopped to release the energy stored in the inductor by means of an auxiliary switch K_a that is necessarily turn-off controlled.

This principle of "dual forced commutation" leads to the dual circuits of those in Figs. 5.10, 5.11 and 5.12. The versions of these circuits implying a variable frequency operation are shown in Fig. 5.20.

Using a switch K_p that is bidirectional with respect to voltage removes part (Fig. 5.20B) or whole of the phase of linearly rising current in the inductor. Nevertheless, the constant current charging of the capacitor still occurs in the circuit of Fig. 5.20B and there still is a strong dependence of the duration of the commutation on the load current.

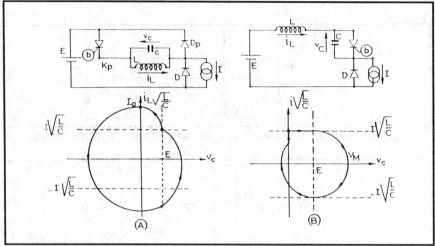

Fig. 5.20 Dual forced commutated chopper with no auxiliary switch: (a) self-
adapting chopper, and (b) quasiresonant bidirectional ZVS chopper

5.4 FORCED COMMUTATION IN DC/AC CONVERTERS

DC/AC converters are classified into two main types by the nature of the DC
source:
• voltage source inverters with current-reversible switches, the commutating
quantity in each commutation cell of these converters is the current.
• current source inverters with voltage-reversible switches; the commutating
quantity in these converters is the voltage.

Precisely because of the AC operation and in contrast to DC/DC converters,
the different types of forced commutation previously outlined can be applied
directly to DC/AC converters. The circuits for reversing the capacitor polarity or
the inductor current can be dispensed with.

The usual principles of forced commutation with turn-on controlled switches
(Figs. 5.3A and Fig. 5.9A) lead to the circuits of Figs 5.21 and 5.22 (only
DC/single-phase-AC converters are dealt with here).

Nonetheless, it has been shown that these principles also apply to turn-off
controlled switches (Figs 5.23 and 5.24). Note that in these circuits the inductor
is essential to supply energy to the capacitor.

Applying duality rules to the circuits of Fig. 5.21 to 5.24 gives the DC/AC
converters involving the dual forced commutation described in the preceding
paragraphs.

Finally, it is difficult to speak of the forced commutation without referring to
a converter that is very popular in the industrial world. This is the current source
inverter with isolation diodes (Fig. 5.25) that can be modulated without
requiring auxiliary switches.

75

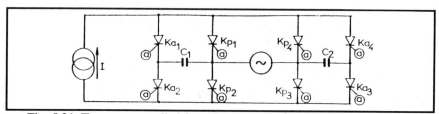

Fig. 5.21 Turn-on controlled forced commutated current-source inverter

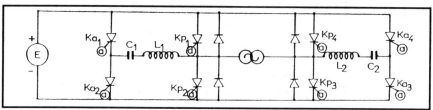

Fig. 5.22 Turn-on controlled forced commutated voltage-source inverter
(Mac-Murray)

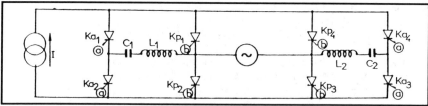

Fig. 5.23 Turn-off controlled forced commutated current-source inverter

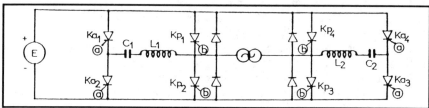

Fig. 5.24 Turn-off controlled forced commutated voltage-source inverter

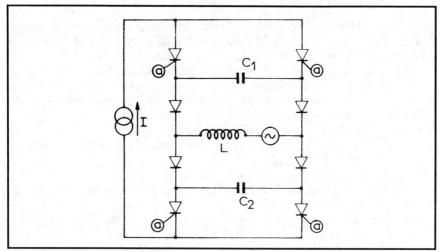

Fig. 5.25 Turn-on controlled forced commutated current-source inverter with isolating diodes (typically using thyristors)

By duality, it is also possible to achieve modulation in a voltage source inverter (with isolation diodes) without any auxiliary switch (Fig. 5.26).

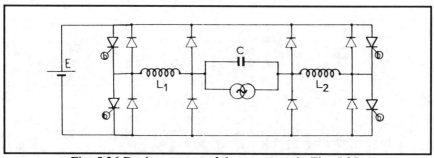

Fig. 5.26 Dual converter of the converter in Fig. 5.25: turn-off controlled voltage-source inverter using, for example, GTOs operating in the dual-thyristor mode.

With this voltage inverter, all the safety of operation of the current source inverter is preserved. The turn-off controlled switches with inherent turn on under zero voltage prevent any shorting of the voltage source. Furthermore, this technique is totally compatible with a half-bridge structure (a single inverter leg).

5.5 QUASI-RESONANT CONVERTERS

Nowadays, a new generation of converters is arousing considerable interest among power electronics designers looking for higher and higher frequencies of operation to improve the power density and response times of the circuits.

These so-called quasi-resonant converters achieve DC/DC conversion and are characterized by the use of reactive components that shape the current and/or voltage into quasi-sinusoidal discontinuous waves, thus reducing the commutation losses in the switches as these either turn on at the voltage zero crossing or turn off at the current zero crossing [56~61].

Hence quasi-resonant converters basically operate in the soft commutation mode. They must be composed of:
• at least one controlled device to regulate the power flow, and
• auxiliary circuits that provide conditions for inherent commutation of this switching device.

Obviously, the operation of these quasi-resonant converters is based on the forced commutation principle in the way it has been described previously.

The synthetic approach that has been carried out leads to the four single-switch quasi-resonant converters in Fig. 5.27.

The quasi-resonant converters using the dual forced commutation principle are grouped in Fig. 5.28. In contrast to the circuits in Fig. 5.27, the energy is stored in an inductor and main switch K_p is turn-off controlled.

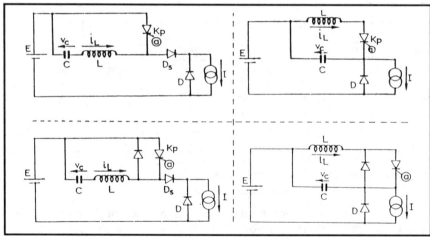

Fig. 5.27 Zero current switching quasi-resonant chopper

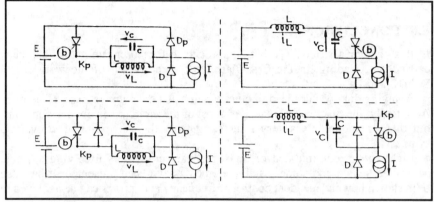

Fig. 5.28 Zero voltage switching quasi-resonant chopper

In passing, it should be noted that in each of the basic circuits of Fig. 5.27 and 5.28, variations are possible by moving the positive terminal of the capacitor from the negative pole of source E to its positive pole and/or by moving current source I from downstream of inductor L to upstream of inductor L. Such modifications only introduce an extra DC component in the voltage across the capacitor and/or in the inductor current.

The study of all these circuits was carried out in a thesis [62].

In these quasi-resonant converters, for a given input voltage E, and for given L and C values, the operation of the converter is fully determined by the value of the load current I. Assuming the latter to be constant, since the energy is delivered in a discontinuous way the power supplied to current source I is directly proportional to the frequency, and so is the voltage across the current source.

Without losing the soft commutation, independent control of the frequency and the power in quasi-resonant converters can be restored by freezing the capacitor voltage or the inductor current during the conduction phase of main switch K_p. This can be achieved by means of a second controlled switch that also operates in the soft commutation mode. Furthermore, this freezing should result in an inherent turn on or turn off of this auxiliary switch since its controlled commutation is intended to release the resonant network from this frozen phase.

The evolution of the voltage across the capacitor can be frozen in two different ways:

1 . when the stored energy reaches its peak, at the zero crossing of the current flowing through it, by means of a switch with inherent turn off, and
2 . when the voltage reaches a predetermined voltage, generally either 0 or E, by means of a switch with inherent turn on.

Dual processes are used to freeze the current in the inductor.

5.6 CONCLUSION

In this chapter a strictly synthetic approach to forced commutation has been carried out, leading, in the first place, to the recreation of the well-known DC/DC and DC/AC circuits.

Applying duality rules lead to other circuits that are more or less well-known. Admittedly, the practical relevance of most of these circuits is not very obvious, but they are nonetheless very interesting from the viewpoint of converters synthesis.

Eight forced commutated or quasi-resonant converters involving a single controlled switch and some reactive elements have been presented. It has also been shown how independent control of frequency and power can be achieved in these converters.

Forced commutation has long been the only recourse when the only power switches available were the thyristor and the diode, which can naturally only operate in the soft commutation mode. The size, weight and cost of the forced commutation circuits were major problems in the design of equipment, the actual size, etc., being related to the turn-off speed of the thyristors.

Though commutation problems are solved with totally controlled devices, these circuits may again become of practical value, for three major reasons:

1 . The zero current commutation (thyristor) and the zero voltage commutation (dual thyristor) are lossless.

2 . The waveforms are quasi sine-waves, with all the advantages that follow from this.

3 . The commutation circuit benefits from the size (and price) reduction following from the frequency increase allowed by faster switches.

Under these circumstances, a number of authors think that quasi-resonant converters are a more favourable compromise than the switch-mode versions.

Thinking at the switch level, an extra point can be made concerning the function of the auxiliary reactive circuit. If the control permits turn on and turn off independently of the state of the switch, the auxiliary circuit can be considered as a snubber that enables these commutations to be performed in the most favourable conditions, turn off at the zero crossing of the current, and turn on at the zero crossing of the voltage. By contrast, if the control depends directly on the voltage across the switch or on the current flowing through the switch (inherent commutation), which naturally occurs in the thyristor and is artificially implemented in the dual thyristor, the auxiliary circuit is then absolutely necessary to force this inherent commutation.

6

STATIC CONVERSION
THROUGH AN AC LINK

6.1 INTRODUCTION

The diagram of such an energy conversion system between two electrical systems S_1 and S_2, suitable for the supply or absorption of electrical energy, is shown in its general form in Fig. 6.1. The converters CS_1 and CS_2 control the energy flow between the systems S_1 and S_2 on the one hand, and the intermediate AC stage S_i on the other hand. Depending on the nature of systems S_1 and S_2, which may be DC sources or single or polyphase AC sources, the static converters CS_1 and CS_2 can be DC/AC inverters or AC/AC converters.

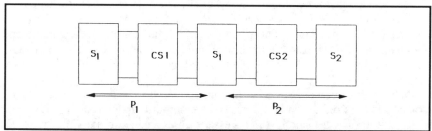

Fig. 6.1 Principle of AC link conversion

Assuming the intermediate stage S_i is single phase and capable of either supplying or absorbing power, this energy conversion arrangement is of no particular novelty since it basically consists of two sources S_1 and S_2 that independently exchange powers P_1 and P_2 with a third system S_i. An example of such a system would be the single phase mains supplying two DC or AC loads S_1 and S_2.

By contrast, if the source S_1 supplies power P_1 while S_2 absorbs power P_2 equal to P_1 (Fig. 6.2), there is no net power flow into or out of S_i and it can therefore consist simply of a network of inductors (possibly coupled) and capacitors, or even be replaced with a direct connection between the converters CS_1 and CS_2 [52,53].

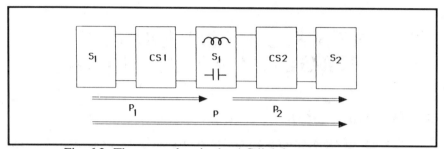

Fig. 6.2 The power lost in the AC link is zero: $P_1=P_2=P$

In this chapter all types of electrical energy conversion will be realized using the scheme in Fig. 6.2 under the following two conditions:

1 . Soft commutation operation for converters CS_1 and CS_2, which implies that their frequencies must be equal to that of the AC source.

2 . Source S_1 is a voltage source, since all the results obtained with a voltage source can be transformed to the case where S1 is a current source by duality.

The conversion principle based on the scheme in Fig. 6.2 — in the which converters CS_1 and CS_2, operating in the soft commutation mode, can utilize a high frequency AC link — is of utmost interest in the present development of power electronics in view of the trend towards size, weight and noise reduction, but also towards efficiency and reliability improvements.

It should also be noted that in all these conversion processes, when galvanic isolation is required or when the ratio of the output voltage to the input voltage is very large (or very small), the necessary transformer can naturally be placed within this high frequency link with all the properties that it entails.

However, in order to keep this presentation simple, the transformer will be disregarded wherever possible.

6.2 DC/DC CONVERSION

6.2.1 Basic rules

Power generally flows from DC source S_1 to DC source S_2. Thus CS_1 usually operates in the inverter mode while CS2 operates in the rectifier mode.

Assuming S_1 to be a voltage source (E_1), the structure of the conversion chain depends fundamentally on the type of load, i.e. whether:

• S_2 is a current source (I_2), or

• S_2 is a voltage source (E_2)

In the case of **DC voltage/DC current conversion**, converters CS_1 and CS_2 can be directly connected to each other (Fig. 6.3).

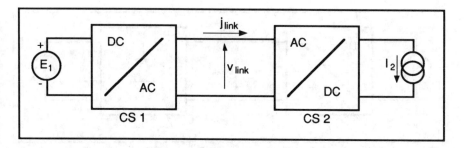

Fig. 6.3 AC link direct "DC voltage/DC current" converter

In this topology, which has already been discussed in Chapter 4, converter CS_1 fixes the amplitude and the frequency of the AC voltage v_{link} (self-commutating inverter) and converter CS_2 operates in the voltage rectifier mode [20,31,45].

Soft commutation of converter CS_1 (Fig. 6.3) results in bidirectional power flow in source E_1. Thus converter CS_2 must also allow bidirectional power flow in current source I_2, which means that it must be fully controlled.

It has been shown (section 4.4.3) that the switches of converters CS_1 and CS_2 had to have dual commutation mechanisms depending on the relative phase-shift between v_{link} and j_{link}. Furthermore, this phase shift is the parameter that controls the power flow.

Whatever commutation mechanisms are used, the input and output waveforms are identical to those observed in a chopper with reverse voltage capability (asymmetrical half-bridge) operating in the constant frequency PWM mode. However, the indirect conversion offers the potential advantage of galvanic isolation without any duty cycle limitation.

Nevertheless, in many applications, there are various reasons which lead to the switches in CS_1 and CS_2 being chosen to have the same commutation mechanisms. Under such circumstances, it may be necessary to employ some ingenious strategy to enable CS_1 to operate in the soft commutation mode.

In a similar fashion, if reverse power flow is not required, the converter can be greatly simplified by replacing the fully controlled converter CS_2 with a diode bridge (Fig. 6.4). As a consequence of this simplification, it will be necessary to arrange for the soft commutation of inverter CS_1.

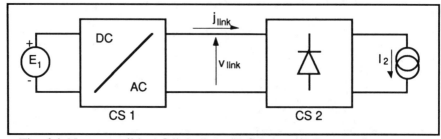

Fig. 6.4 Non reversible AC link direct "DC voltage/DC current" converter

In the case of **DC voltage/DC voltage conversion,** an element that behaves like a current source must be provided in the AC link (Fig. 6.5) in order to comply with the fundamental laws governing the interconnection of sources.

An energy-storage element simultaneously able to exchange energy with either E_1 or E_2 is thus arranged in this manner. By definition, the current in current source j_{link} cannot change instantaneously; this implies that a commutation of CS_1 (respectively CS_2) always causes a change of sign of the current and thus of the power delivered by E_1(respectively E_2). The sign of the current at the commutation time determines the commutation mechanisms of the switches of CS_1 and CS_2.

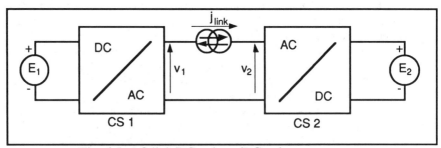

Fig. 6.5 AC link DC voltage/DC voltage converter

Converter CS2 in Fig. 6.5 may or may not be reversible.

In the following section, we will deal with several principles allowing soft commutation in DC voltage/DC current conversion, and then in DC voltage-DC voltage conversion, but we will mainly focus on the non-reversible topologies and will only briefly mention the specific features of reversible converters.

6.2.2 DC voltage / DC current conversion

(a) Solution 1

A classic technique, used particularly with thyristors [49,63], involves placing a current source in parallel with converter CS_2. It is assumed that this current source cannot be a source or sink of real power. The corresponding topology is shown in Fig. 6.6 with the associated waveforms in the highly idealized case of a sinusoidal current source and with converter CS_2 being a diode rectifier.

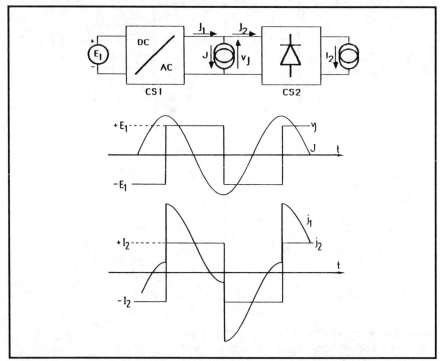

Fig. 6.6 Non-reversible DC voltage/DC current converter and corresponding waveforms (the current source J gives the conditions for soft commutation of the turn-on controlled DC/AC converter CS1)

When the switches of CS_1 are turn-on controlled and turn off inherently, the soft commutation of the inverter is obtained practically by using an LC resonant circuit for current source J, the natural frequency being higher than the operating frequency of CS_1. A variation consists of connecting two back-to-back turn-on controlled switches such as thyristors in series with the LC network [49].

A very similar principle can be used when the switches of CS_1 are turn-off controlled (Fig. 6.7). The current source, which has $\pi/2$ lagging phase with respect to the fundamental of voltage v_j , consists of an LC network with a natural frequency lower than the switching frequency of CS_1, or more simply just of an inductor [45,48]. It is also possible to connect two back-to-back thyristors in series with these networks.

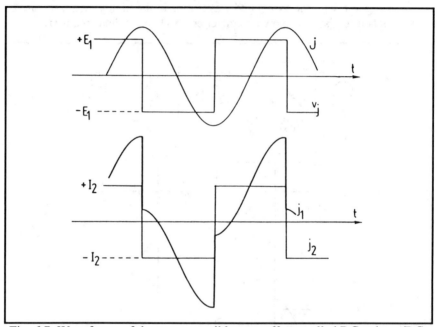

Fig. 6.7 Waveforms of the non-reversible turn-off controlled DC voltage/DC current converter

The major drawbacks of this first solution are:
1 . the current overrating of the switches, and
2 . the complete inability to control power flow.

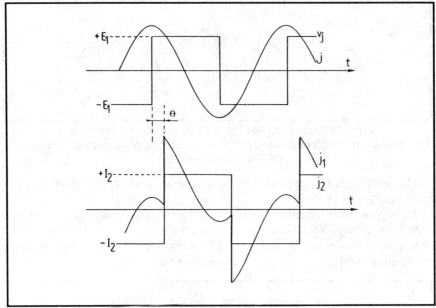

Fig. 6.8 Waveforms of the reversible DC voltage/DC current converter
(the inverter and the rectifier are turn-on controlled)

This first technique can nevertheless be very widely applied when CS_2 is a diode bridge or a controlled rectifier with switches having the same type of commutations as those of CS_1, provided only that the following condition is obtained:

$$I_2 < J_M \cos \theta \qquad (6.1)$$

with θ being defined in Fig. 6.8, and J_M the amplitude of the current source.

(b) Solution 2
On the assumption that the switches of CS_1 are turn-off controlled, the soft commutation mode is possible only if the sign of the current j_1 remains the same before and after the commutation of CS_1. This can be achieved simply by connecting in series with CS_1 an element that acts as a current 'memory'.

Another solution consists of connecting CS_1 and CS_2 via an inductor L as shown in Fig. 6.9.

87

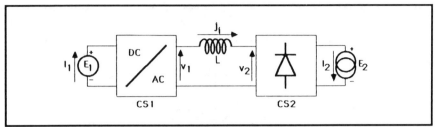

Fig. 6.9 Non-reversible DC voltage/DC current converter
(inductor L gives the conditions for soft commutation of the turn-off controlled
DC/AC converter CS$_1$)

The waveforms in Fig. 6.10 show the operation of this circuit when CS$_2$ is a diode rectifier.

Throughout the current reversal in inductor L, the voltage across current source I$_2$ is zero. This situation is comparable to the well-known overlap phenomenon that occurs in rectifiers. For a given voltage E$_1$ and inductance L, the duration of this zero-voltage plateau depends only on I$_2$.

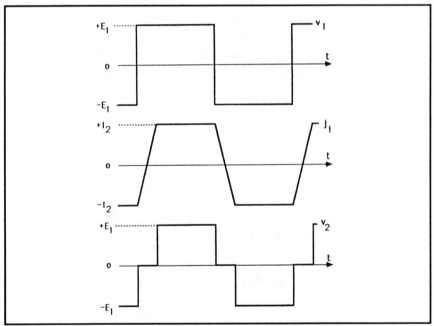

Fig. 6.10 Waveforms of the converter in Fig. 6.9

Consequently, the average voltage E$_2$ across I$_2$ depends on I$_2$ and the switching frequency f$_s$ according to the relation:

$$I_2 = \frac{E_1}{4Lf_s}\left[1 - \frac{E_2}{E_1}\right]$$

(6.2)

Thus the converter behaves as a DC voltage generator that has a non-dissipative internal resistance with a value of $4Lf_s$ (Fig. 6.11).

This technique, whilst not intrinsically requiring any excess current rating for the switches, does lead to a reduction in the controllable real power. It also permits, to a certain extent, power control by means of the switching frequency f_s (by variation of the internal resistance). Nevertheless, frequency control only regulates the short-circuit current, the no-load voltage being determined by voltage source E_1.

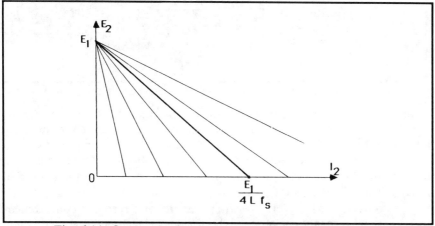

Fig. 6.11 Output characteristics of the converter in Fig. 6.9

Power flow in the circuit of Fig. 6.9 can be made reversible provided CS_2 is fully controlled. However in order to obey the fundamental rules for static converters, the switches of CS_2 have to be turn-on controlled and turn off inherently (thyristors). The converter thus obtained is comparable to that of Fig. 6.3, with inductance L now regarded as a parasitic element (e.g. the leakage inductance of a transformer).

(c) Solution 3

The turn-on controlled soft commutation of inverter CS_1 requires energy to be fed back into voltage source E_1 before each commutation of the inverter. It also means that the current j_1 delivered by inverter CS_1 must change sign before each of its commutations.

With a diode rectifier, this result can only be achieved by inserting an oscillatory (or resonant) network between inverter CS_1 and rectifier CS_2. The arrangement of this network must be chosen to give the best performance. In particular, effective utilization of the switches should be made and the waveforms should be as close to sinewaves as possible in order to limit noise generation. Thus, the resonant network should behave as a low-pass filter as well as complying with the laws governing the interconnection of sources.

One solution is to connect a capacitor in parallel with rectifier CS_2 and to connect an inductor between inverter CS_1 and capacitor C (Fig. 6.12).

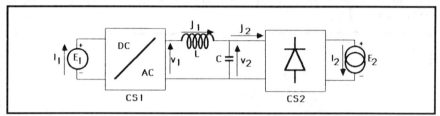

Fig. 6.12 Series-parallel resonant converter

In practice, this last solution amounts to inserting an LC filter in the AC link. However, it should be noted that the purpose of this filter is very particular insofar as it is the phase between the output quantities of the inverter CS_1 i.e. v_1 and j_1 — and consequently the type of commutation — that matters, rather than the quality of the waveform of voltage v_2 delivered by the filter since this is rectified and filtered by CS_2.

The converter of Fig. 6.12 is a first example of a resonant converter: the series-parallel resonant converter. A brief study of the operation of this non-reversible series-parallel resonant converter based on state plane analysis is given in section 6.3.

6.2.3 DC voltage/DC voltage conversion

The arrangement of an AC link non-reversible converter controlling the power flowing between two DC sources is shown in Fig. 6.13. The waveforms corresponding to the operation of this converter when the switches of CS_1 are turn-off controlled are drawn on this same figure assuming current source j to be sinusoidal.

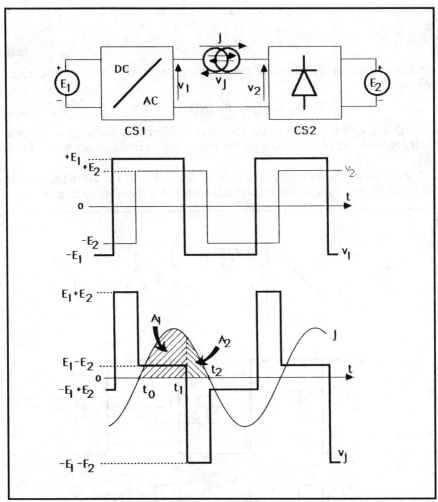

Fig. 6.13 AC link DC voltage/DC voltage converter
topology and waveforms

Let us define A_1 and A_2 such that:

$$A_1 = \int_{t_0}^{t_1} j \, dt$$

(6.3)

and

$$A_2 = \int_{t_1}^{t_2} j \, dt$$

$$(6.4)$$

Since current source j does not sink or source any energy over a half-period, we have:

$$E_1 \, (A_1 - A_2) = E_2 \, (A_1 + A_2) \qquad (6.5)$$

This relation also means that the converter is necessarily a voltage step-down converter, and this is true whatever the commutation mode of the switches of CS_1.

The simplest way to obtain this current source j is to use an inductance L that could be, for example, the leakage inductance of a transformer (Fig. 6.14).

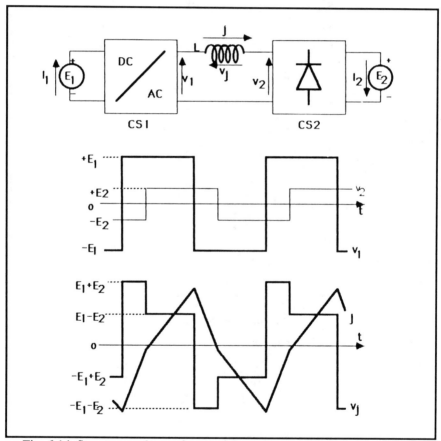

Fig. 6.14 Structure and waveforms of the non-reversible DC voltage/DC voltage converter with inductor L playing the part of AC current J

The analysis of the waveforms in Fig. 6.14 show that the switches of inverter CS_1 must be turn-off controlled (dual thyristors).

Current I_2 is the average value of the full-wave rectified current j. The output quantities E_2 and I_2 are related according to the following equation:

$$I_2 = \frac{E_1}{8Lf_s} \left[1 - \left(\frac{E_2}{E_1}\right)^2 \right]$$

(6.6)

The $E_2(I_2)$ characteristics are shown in Fig. 6.15 for several values of f_s.

This converter therefore behaves as a DC voltage generator with an internal resistance that is non-dissipative and depends on the switching frequency f_s, on inductance L and on the load (E_2,I_2).

From the set of $E_2(I_2)$ characteristics, it can clearly be seen that when E_1 is kept constant the only possible method of power control is to vary the Lf_s product, which amounts to varying f_s (the value of L cannot be varied easily). This variation of f_s leads to a variation of the short circuit current but the no load voltage remains constant and equal to E_1.

In a circuit such as that of Fig. 6.9, power control based on frequency, though very flexible is in fact limited for two major reasons:

1 . The frequency range itself is limited: by transformer saturation at the lower frequencies, and by switching losses at the higher frequencies.

2 . For a constant value E_2 of the output voltage, and especially in the short circuit mode, current I_2 varies as $1/f_s$.

In addition to this narrow range of power control, this converter, which can only operate with turn-off controlled switches, also suffers from a large current overrating required for the switches (with a 2:1 ratio) because of the poor power factor resulting mainly from the inductance in the AC link.

Nevertheless, it is possible to ameliorate several of these problems considerably. In effect, the rectifier-filter-load network always absorbs real power and can in this way be regarded as a resistor (Fig. 6.14). In the light of what was written in Chapter 3, the addition of a capacitor C in series with the inductor allows on the one hand the commutation mode of inverter CS_1 to be chosen and on the other hand the power factor of this same CS_1 inverter to be improved.

Furthermore, increasing the power control range ultimately comes back to increasing the variation of the short circuit current for a given variation of frequency f_s (Fig. 6.15). This is obtained by replacing impedance $L\omega_s$ with an equivalent impedance that varies more quickly than frequency f_s, for example the impedance of an LC resonant network operating in the vicinity of its natural frequency.

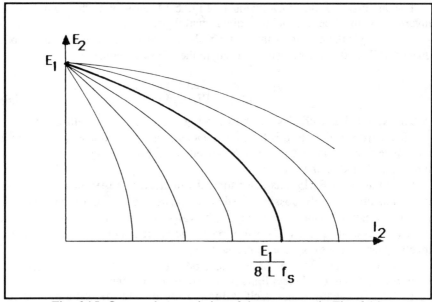

Fig. 6.15 Output characteristics of the converter in Fig. 6.14

Thus, a new resonant converter is obtained (Fig. 6.16) which is known as the series resonant converter.

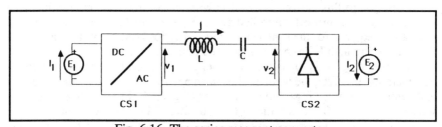

Fig. 6.16 The series resonant converter

6.2.4 Conclusion

In this section devoted mainly to non-reversible DC/DC converters with an AC link, it has been shown that the commutation problems could be solved in the case of, first, a load of the current-source type, and second, a load of the voltage-source type.

Several solutions have been reviewed, all of them capable of being transformed to reversible DC/DC converters. Two of them, the series-parallel and the series resonant inverter stand out because of their general properties which have already been outlined in Chapter 3.

The following section provides a deeper analysis of these two converters including the actual waveforms, and a comparative study of the characteristics inherent in their topologies.

6.3 RESONANT CONVERTERS

6.3.1 The series resonant converter

Assuming the DC voltage sources to be perfectly filtered and the switchings to be instantaneous, study of the series resonant converter resolves to study of the response of an assumed lossless series resonant network connected on one side to a voltage source $v_1 = \pm E_1$ of frequency f_s, and on the other side to a voltage source $v_2 = \pm E_2$ in phase with the current in the resonant network (Fig. 6.17).

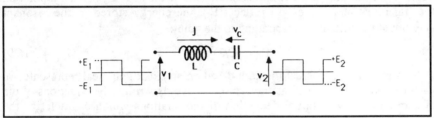

Fig. 6.17 Equivalent circuit for the analysis of the series resonant converter

Under these conditions, unlike in the series resonant inverter, the current in the resonant network of the series resonant converter is necessarily oscillatory. The frequency of these oscillations is f_0, the natural frequency of the LC resonant network.

Limiting the study to short turn-on or turn-off control pulses, three modes of operation can be distinguished:

1. $f_s < f_0/2$: discontinuous conduction mode. All the commutations of inverter CS_1 occur at the current zero-crossing in the resonant network.

2. $f_0/2 < f_s < f_0$: continuous conduction mode.

3. $f_0 < f_s$: continuous conduction mode.

In the first two modes, with the switching frequency less than the natural frequency, the switches of inverter CS_1 are turn-on controlled and turn off inherently (thyristors). In the third mode, the switches of inverter CS_1 are turn-off controlled and turn on inherently (dual thyristors).

(a) Method of study

Given the very simple circuit resulting from the functional representation of inverter CS_1 and the rectifier-filter-load network, an analytic study can usefully be carried out [64].

The response of the circuit is well known and the equations giving current j and voltage v_C for one mode are:

$$v_C = V_{LC} - (V_{LC}-V_{C0}) \cos \omega_0 t + (J_0/C\omega_0) \sin \omega_0 t \qquad (6.7)$$

$$j = C\omega_0 (V_{LC}-V_{C0}) \sin \omega_0 t + J_0 \cos \omega_0 t \qquad (6.8)$$

in the which V_{LC}, V_0, J_0 and ω_0 are respectively the voltage applied across the resonant network, the initial conditions for the mode being considered and the natural angular frequency of the LC network.

This study is greatly facilitated by using the state plane representation $(v_C, j\sqrt{L/C})$ [20~23;65~68]. In this plane, the trajectory of the point of operation is an arc of a circle centred at the point with coordinates $(v_{LC},0)$ starting at the point corresponding to the initial conditions. The distance d between a point on the trajectory and the origin reflects E_n, the energy stored in the resonant network at the instant corresponding to the point:

$$E_n = (C/2) d^2 \qquad (6.9)$$

This analytico-graphical method, based on an exact graphical representation — for steady state as well as transiently — avoids writing the equations of the system and allows straight deduction of the relations which characterize the operation of the mode being studied.

This method of study, already widely used in thyristor forced-commutated circuits [50] or for input and output filters, is a very efficient tool for studying the resonant converters.

(b) Discontinuous conduction mode $(f_s<f_0/2)$
The state-plane analysis of the discontinuous conduction mode is given in Fig. 6.18. M_1 and M_2 represent discontinuous conduction during which voltage v_c and current j are frozen.

It can be readily inferred from the state plane in Fig. 6.18 that the peak voltage across the capacitor is independent of the operating conditions and is:

$$V_{cmax} = 2 E_1 \qquad (6.10)$$

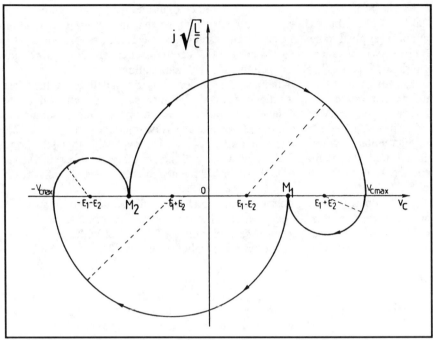

Fig. 6.18 State plane analysis of the discontinuous mode of the series resonant converter

The peak current in the switches can also be deduced directly from the state plane:

$$I_{max} = \left(1 + \frac{E_2}{E_1}\right) E_1 \sqrt{\frac{C}{L}}$$

(6.11)

and this current increases with E_2/E_1.

As voltage E_2 tends to E_1, the amplitude of the current flowing in the diodes of inverter CS_1 decreases. When $E_2 = E_1$, this current vanishes and the diodes stop conducting.

Current I_2 equals the average value of the full wave rectified current j:

$$I_2 = \frac{4}{\pi} \frac{f_s}{f_0} E_1 \sqrt{\frac{C}{L}}$$

(6.12)

This current is independent of E_2 and can be expressed exclusively as a function of f_s/f_0 and of the constant parameters of the circuit.

Thus, in the discontinuous mode, the series resonant converter behaves as a frequency controlled average current generator. However, it suffers from several drawbacks:

97

1 . The form factor of the current in the resonant network is very poor. The peak current in the switches is thus very high so that this mode of operation cannot be applied to switches rated by instantaneous current (transistors,....). By contrast, switches rated by RMS current that can withstand significant overloads (thyristor, GTO) [69~71] are suitable for such applications.

2 . Though all the commutations occur under zero current and do not therefore suffer from problems due to diode recovery, the turn on of a switch is liable to generate a severe dV/dt on the complementary switch of the same commutation cell. The amplitude of this dV/dt can cause unwanted conduction of this complementary switch. This phenomenon is well known with thyristors, and often limits the switching frequency because of the auxiliary (snubber) circuits it requires. In this case, the GTO thyristor has the undeniable advantage of turn-off control [71] .

3 . The energy transfer control narrow range since the output current varies in the same ratio as the frequency. A solution commonly used to solve this problem consists in connecting a second inductor in parallel with the capacitor to obtain a network with both series and parallel resonance [71].

(c) Continuous conduction mode ($f_o/2 < f_s < f_o$)
The study of this mode of operation has already been well developed in the literature [5~16;72~76] and we will only review the state plane (Fig. 6.19) and a few characteristic variables such as the peak voltage across the capacitor:

$$V_{Cmax} = \frac{(1+q)(1- \cos \theta_D)}{q- \cos \theta_D} E_1$$

(6.13)

and the peak current in the resonant network:

$$I_{max} = \frac{1+q^2- 2 \cos \theta_D}{q- \cos \theta_D} E_1 \sqrt{\frac{C}{L}}$$

(6.14)

with $q = E_2/E_1$.

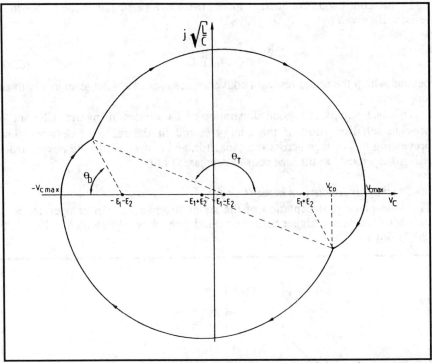

Fig. 6.19 State plane analysis of the continuous mode of the series resonant
converter operating below the natural frequency

The amplitude of the voltage across the capacitor at the instant of commutation is given by:

$$LV_{c0} = q\, V_{Cmax} \tag{6.15}$$

and the average current I_2 in the load by:

$$I_2 = \frac{2}{\pi} \frac{V_{Cmax}}{E_1} \frac{f_s}{f_0} E_1 \sqrt{\frac{C}{L}} \tag{6.16}$$

with

$$f_s/f_0 = \pi /(\theta_T + \theta_D) \tag{6.17}$$

According to the state-plane analysis of Fig. 6.19, the diodes of the inverter cannot conduct any current unless the following condition is satisfied:

$$V_{cmax} > E_1 + E_2 \tag{6.18}$$

On the other hand, if the diodes stop conducting, we necessarily have:

$$E_1 = E_2 \tag{6.19}$$

In this continuous conduction mode, there is a particular value of the load resistor R_C given by:

$$R_c = \frac{\pi}{4} \frac{f_0}{f_s} \sqrt{\frac{L}{C}}$$

(6.20)

beyond which the series resonant converter behaves as a voltage source with an amplitude E_1.

This last remark is a good illustration of the diodes in inverter CS_1 act to provide self-limitation of the energy stored in the resonant network thus preventing the voltage across the capacitor, and consequently the power, from rising dangerously as the load resistor decreases [4,6].

(d) Continuous conduction mode ($f_o < f_s$)

The state-plane representation of the series resonant converter operated at a switching frequency higher than the natural frequency is presented in Fig. 6.20 [20,21,66].

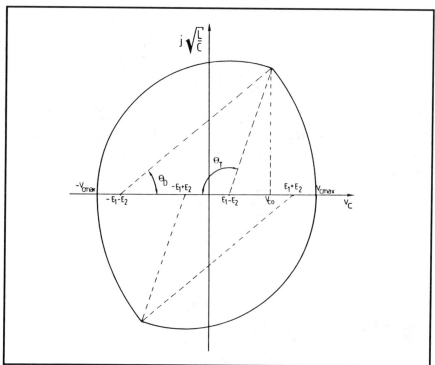

Fig. 6.20 State plane analysis of the continuous mode of the series resonant converter operating above the natural frequency

The peak voltage across the capacitor is

$$V_{Cmax} = \frac{(1-q)(1-\cos\theta_T)}{q + \cos\theta_T} E_1$$

(6.21)

with $q = E_2/E_1$.

The amplitude of the voltage across the capacitor at the instant of commutation is always given by:

$$V_{c0} = q\, V_{Cmax}$$

(6.22)

and I_2, the average current in the load, remains as:

$$I_2 = \frac{2}{\pi} \frac{V_{Cmax}}{E_1} \frac{f_s}{f_0} E_1 \sqrt{\frac{C}{L}}$$

(6.23)

with

$$f_s/f_0 = \pi /(\theta_T + \theta_D)$$

(6.24)

(e) Output characteristics $E_2(I_2)$

The set of expressions established in this paragraph 6.3.1 enable plotting of the $E_2(I_2)$ characteristics of the series resonant converter in the various modes of operation taking f_s (or the f_s/f_0 ratio) as a control parameter (Fig. 6.21).

Above the natural frequency, the $E_2(I_2)$ characteristics are very close to those obtained from the calculations based solely on the terms involving the fundamental, since the current harmonics are clearly more attenuated in this case.

These static characteristics are very important since they enable determination of the power range available in relation to a certain frequency band and also show whether limiting the control parameter range limits the output current or whether auxiliary limiting circuits are required.

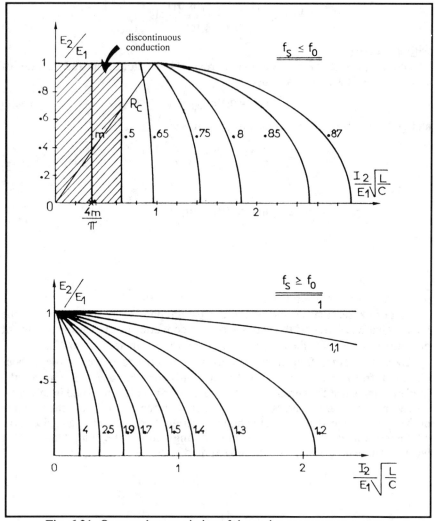

Fig. 6.21 Output characteristics of the series resonant converter

6.3.2 The series-parallel resonant converter

Assuming the DC current and DC voltage sources are perfectly filtered and the commutations are instantaneous, the study of the series-parallel resonant converter [67,77~81] reduces to that of a supposedly lossless LC resonant network (Fig. 6.22) connected, on the one hand to a voltage source $v_1 = \pm E_1$ with frequency f_s, and on the other hand to a current source j_2 with amplitude $\pm I_2$ such that:

102

$$j_2 = I_2 * \text{sgn}(v_2) \qquad\qquad (6.25)$$

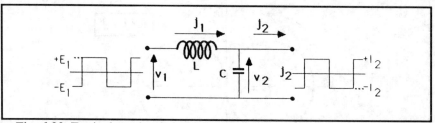

Fig. 6.22 Equivalent circuit for the analysis of the series-parallel resonant converter

In the circuit of Fig. 6.22, there are many modes of operation depending on both the frequency f_s and the amplitude of the load current I_2. The presentation and analysis of these numerous modes is greatly facilitated, as previously, by the use of state-plane representation.

In this section, only a qualitative description of these modes of operation will be given, assuming the control signals of the switches to be short turn-on or turn-off pulses.

(a) No-load operation ($I_2 = 0$)
The no-load operation is in every respect analogous to the short circuit operation of a series resonant converter. The state plane representation of the various possible cases is given inFig. 6.23.

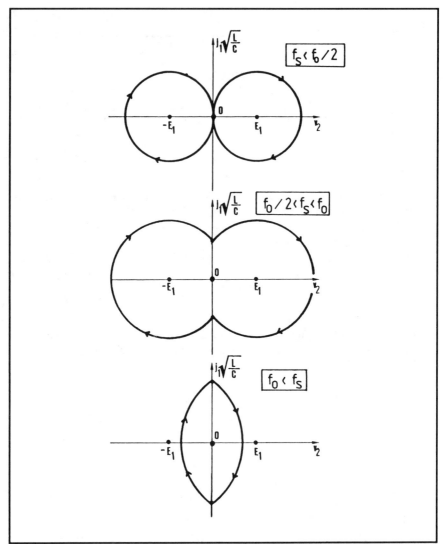

Fig. 6.23 No-load operation of the series-parallel resonant converter

If $f_s < f_0/2$, the conduction is discontinuous and after a complete period of oscillation current j_1 vanishes and remains zero for a time depending on the control. The commutation must be turn-on controlled if switching frequency f_s is less than natural frequency f_0 and it must be turn-off controlled in the opposite case.

(b) Short circuit operation ($E_2 = 0$)

In this mode, voltage v_2 across the capacitor voltage must be zero at each instant so that current I_2 must flow through the rectifier diodes which are all conducting simultaneously.

Current j_1, that results from the application of voltage v_1 across inductor L, is triangular with an amplitude of $E_1/4Lf_s$ and the commutations of inverter CS_1 must be turn-off controlled. In the general case when current source I_2 is only made up of a smoothing inductor, its amplitude is also $E_1/4Lf_s$.

(c) Remark

From the analysis of the no-load operation and of the short circuit mode, it can be inferred immediately that when the switching frequency is less than the natural frequency, the commutation mechanisms change as the load varies. On the contrary, for switching frequencies greater than the natural frequency, the commutation are a priori turn-off controlled whatever the load value.

(d) Intermediate load operation ($I_2 < E_1 \sqrt{C/L}$)

Starting from the no-load operation in the discontinuous mode, current I_2 is increased. In the first instance, the study is limited to I_2 constant and less than $E_1 \sqrt{C/L}$.

During the discontinuous conduction, represented by point 0 in Fig. 6.24, current I_2 is free-wheeling and loops through the rectifier. The turn-on control of the switches of inverter CS_1 leads to a linear increase of current j_1 in inductor L until $j_1 = \pm I_2$.

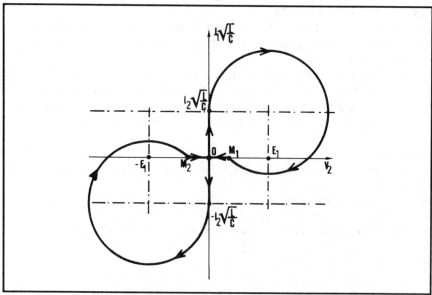

Fig. 6.24 First mode of operation

Then, after each conduction of a switch in CS_1, current j_1 vanishes and causes the aforementioned switch to turn off (Fig. 6.24, points M_1 and M_2). Then capacitor C undergoes a constant current discharge at the end of which current I_2 free-wheels again through the rectifier bridge (Fig. 6.24, point 0).

In this first mode of operation (Mode 1), after each half-period, the resonant network recovers its initial state. An asymmetrical structure like that of Fig. 6.25 can then be considered. We find again here the structure of the quasi-resonant converter presented previously (Fig. 5.14).

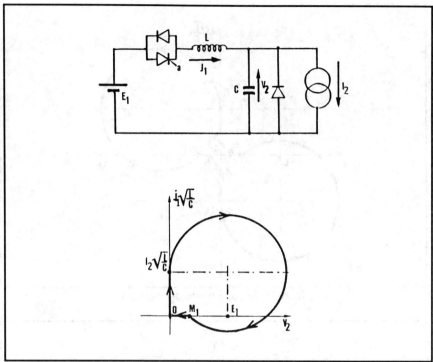

Fig. 6.25 Bidirectional quasiresonant ZCS converter

Increasing the switching frequency leads to a second mode of operation (Mode 2) very similar to the former one. The switches in CS_1 are turned on before the end of the constant current discharge of capacitor C (Fig. 6.26, points M_3 and M_4).

In both these modes, there is no real commutation of the switches of CS_1. The commutation phenomenon only takes place when the switching frequency equals or exceeds the critical frequency at which the turn on of the switches (Fig. 6.26, points M_3 and M_4) coincides with the vanishing of current j_1 (Fig. 6.26, points M_1 and M_2). The representation in the state plane of this mode with real commutations (Mode 3) is given in Fig. 6.27.

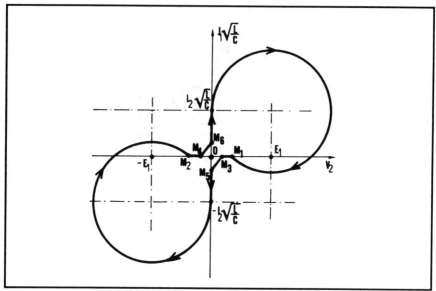

Fig. 6.26 Second mode of operation

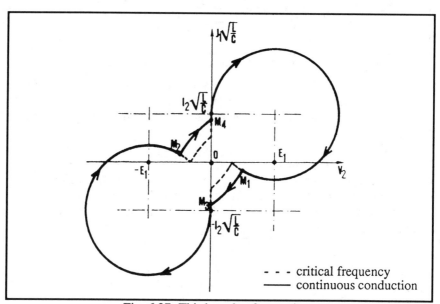

Fig. 6.27 Third mode of operation

Up to now there has always been a free-wheeling period in the rectifier during the which the inductor current rises linearly, this period finishing when $j_1 = \pm I_2$. Always provided I_2 is kept constant, an increase in the switching frequency eliminates this free-wheeling period and leads to the instantaneous commutation of the rectifier. In these circumstances, two modes of operation which mainly differ by the commutation mechanisms of inverter CS_1 — directly determined by the sign of the ordinate of point M_1 — can be obtained:

• negative (Mode 4), the commutations are turn-on controlled (Fig. 6.28) and the switching frequency is less than the natural frequency,

• positive (Mode 5), the commutations are turn-off controlled (Fig. 6.29) and the switching frequency is greater than the natural frequency.

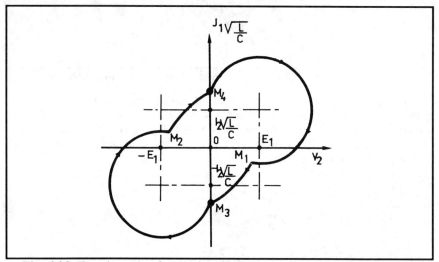

Fig. 6.28 Fourth mode of operation (the switches are turn-on controlled)

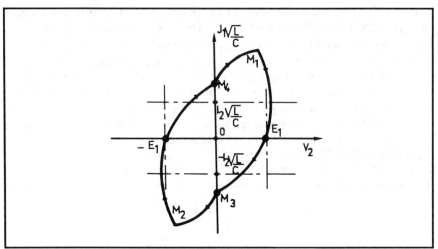

Fig. 6.29 Fifth mode of operation (the switches are turn-off controlled)

Starting from this last mode of operation, if the switching frequency keeps on increasing, the free-wheeling period in the rectifier, with the linear rise of the inductor current, reappears (Mode 6) as shown in Fig. 6.30. This phenomenon is accompanied by a reduction of the voltage across the load and the limit of Mode 6 corresponds to short circuit operation.

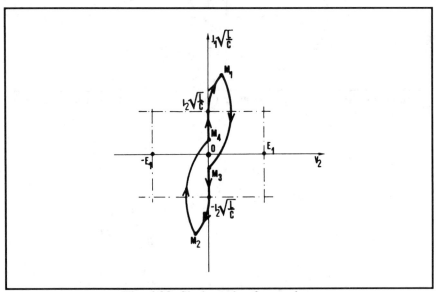

Fig. 6.30 Sixth mode of operation

(e) Intermediate load operation $(I_2 > E_1 \sqrt{C/L})$

When current I_2 is greater than $E_1 \sqrt{C/L}$, the commutations of the switches of the inverter are necessarily turn-off controlled. The representation in the state plane is then similar to:
- that of Fig. 6.29 (Mode 5) if the commutation of the rectifier is instantaneous,
- that of Fig. 6.30 (Mode 6) in the opposite case.

It is important to note that in both these modes the controlled turn off of the switches in CS_1 (Figs. 6.29 and 6.30; points M_1 and M_2) finishes a period with a duration that is determined except for a $k*2\pi \sqrt{LC}$ ($k \in N$) additional term, even in the case of short turn-off pulses.

(f) Output characteristics $E_2(I_2)$

Excluding all other modes that would involve several conductions of the same switch within the same period or use conduction angles for the switch greater than 2π, equations derived from the state planes of Figs 6.23 to 6.30 give the output characteristics $E_2(I_2)$ of the series-parallel resonant converter in the different modes; frequency f_s (or the ratio f_s/f_0) is chosen as a control parameter (Figs 6.31 and 6.32).

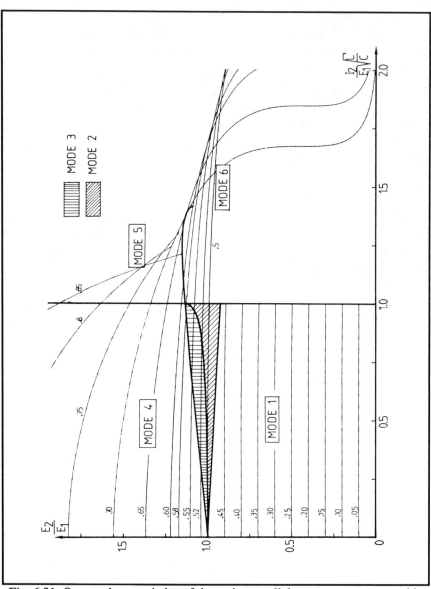

Fig. 6.31 Output characteristics of the series-parallel resonant converter with f_s/f_0 as parameter. (The commutations are controlled by turning on the blocking switches when I_2 is less than $E_1 \sqrt{C/L}$ and by turning off the conducting switches when I_2 is greater than $E_1 \sqrt{C/L}$)

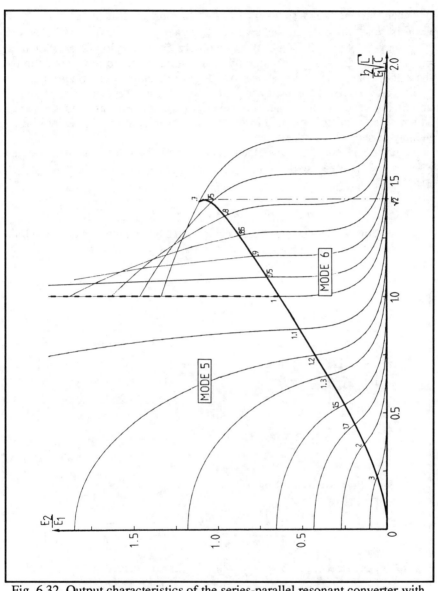

Fig. 6.32 Output characteristics of the series-parallel resonant converter with
f_s/f_0 as parameter. (In this case, the commutations are always controlled by
turning off the conducting switches)

113

When current I_2 is less than $E_1 \sqrt{C/L}$, turn-on control of the switches in CS_1 is possible only if the switching frequency is less than the natural frequency (Modes 1, 2, 3 and 4). Furthermore, in this region of operation, the output characteristics being naturally of the voltage source type, regulation of the output voltage necessitates a narrow range of variation of the frequency.

By contrast, if the frequency is greater than the natural frequency (Modes 5 and 6), the series-parallel resonant converter rather behaves as a current source. The regulation of this output current can however be tricky, especially in the vicinity of short-circuit operation.

Figures 6.31 and 6.32 also show that even if the operation of the series-parallel resonant converter is possible with I_2 greater than $E_1 \sqrt{C/L}$, it is nevertheless difficult to exploit because of its distorted characteristics.

6.3.3 Structural properties of resonant converters

In this section, we propose to make a comparative and qualitative analysis of the properties inherent in the structures of the series and series-parallel resonant inverters reviewed in Fig. 6.33.

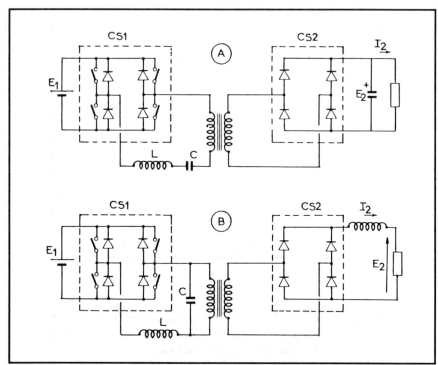

Fig. 6.33 Structures of the series resonnant converter (a)
and series-parallel resonant converter (b)

These two converters differ mainly in the operation of their rectifiers CS_2 and the very different constraints under which they operate, and by the consequences of including a transformer.

In the series converter (Fig. 6.33a), the commutation of the diodes occurs at the zero crossing of the current with a dI/dt that is naturally limited by the resonant network. The switching losses are thus minimized in the diodes. Since no energy is stored in the stray inductance at the time of the commutation, the diodes do not suffer any overvoltage.

In the structure of Fig. 6.33b, the constraints are different depending on the position of capacitor C. If the capacitor is on the secondary side of the transformer, the leakage inductance is 'hidden' in the inductance of the resonant network and the dI/dt in the diodes during the commutations is poorly controlled It can be very high and result in reverse diode currents that, in turn, induce voltage overshoots. So these diodes must generally be equipped with snubbers. By contrast, if capacitor C is on the primary side of the transformer, the leakage inductance results in non-negligible switching times and hence a volt-drop due to the commutation.

As far as the transformer is concerned, in the series converter of Fig. 6.33a, capacitor C is series connected with the transformer thus preventing any DC current from flowing through the transformer. In this series converter the leakage inductance is part of the inductance L of the resonant network and thus plays an integral part in the operation of the converter. The design of the transformer, which may have a very high or very low turns ratio is then made easier because the coupling of the windings can be quite loose which gives very low inter-windings capacitance.

Finally, in the series structure, if CS_1 is a half bridge, resonant capacitor C can also act as a voltage divider.

6.3.4 Conclusion

With turn-on controlled switches such as thyristors, the series-parallel resonant converter has the undeniable advantage of the natural voltage source type characteristics. However, in addition to the problems related to the implementation of inverter CS_1 with such switches, current I_2 must necessarily be limited to a value less than $E_1 \sqrt{C/L}$.

Because of the favourable commutation conditions in rectifier CS_2, the series resonant converter has a structure that is most certainly more suitable for high frequency operation. This is all the more significant when reverse power flow is considered since rectifier CS_2 then becomes a voltage-source inverter in the case of the series resonant converter while it becomes a current-source inverter in the series-parallel resonant converter (c.f. section 3.4).

For many reasons, discussed in previous chapters, and in order to keep the presentation simple, this work is focused on the series resonant inverter of Fig. 6.33a, and more especially on operation above the natural frequency.

We will go thoroughly into the analysis by studying successively:

• The variation of the output voltage in relation to low frequency (compared to the switching frequency) small signal perturbations of the control signal or of the supply voltage E_1 (Appendix 2) [82~87].

• The effect of the capacitive snubbers in the inverter and rectifier (Appendix 3) [66].

Nevertheless, by virtue of the principles of duality, many of the results presented in the following can easily be transposed to the converter of figure 6.33b by any knowledgeable person!

6.4 POWER CONTROL IN RESONANT CONVERTERS

Focusing on the series topology, the different control methods commonly used in such converters are reviewed. Finally, a few control methods that allow constant frequency operation will be presented [88~90].

6.4.1 Variable frequency control

The most widely used control strategies for the resonant converter are:
• Control of the conduction time of the diodes of inverter CS_1. This control is suitable for cases when the commutations of inverter CS_1 are caused by controlled turn on. This control strategy has grown out of the long experience acquired with thyristors for which controlling the reverse bias time is of the utmost importance.
• Control of the conduction time of the controlled switches [87]. This control is only easily implemented when it is a turn off that causes the commutation of inverter CS_1. Thus, this control strategy requires turn-off controlled switches.
• The analog signal to discrete time interval converter" (ASDTIC) control strategy [6]. This has been applied especially to the series resonant converter. The switching times of the switches in inverter CS_1 are determined by the integral of the difference between a voltage proportional to the full-wave rectified current in the resonant network and a reference voltage going to zero. Thus, in practice, the controller effectively regulates the output current to a preset value. Equipped with such a control strategy, the converter behaves as a current source.
• An optimal strategy which, when operating above (respectively below) the natural frequency, amounts to defining trajectories on the phase plane which represent the sequence of inverter diode (respectively transistor) conduction characteristic of the required steady-state mode (Appendix A4). This optimal controller has the interesting property of giving precise control over the current and voltage in the resonant network at all times. Furthermore, the durations of the transients following a step-change in the reference signal or a sudden change in the load voltage (short circuit) are minimized [88].
• Control of the frequency of inverter CS_1. Voltage v_1 is a square wave with an amplitude of $\pm E_1$, the frequency being imposed by the controller. In contrast to the four control strategies already mentioned, no sensor is required in the converter.

All of these power control techniques share variable frequency operation and have two principal drawbacks:
1. Variable frequency operation results in filtering problems and electromagnetic and acoustical noise. The response time of the converter depends on the operating conditions and varies over a large range.

2. The point of operation ($E_2 = 0$; $I_2 = 0$) is difficult to obtain since it theoretically corresponds to a zero frequency if the commutation is turn-on controlled and to an infinite frequency if the commutation is turn-off controlled (Fig. 6.21). The power variation attainable is a result of a compromise between the voltage overshoots experienced by the inductor and the capacitor in the resonant circuit and the allowable frequency range.

Trying to get rid of both these drawbacks leads us to reconsider the frequency variation and control methods described above.

6.4.2 Fixed frequency control

Assuming the time constant of the load to be far greater than the period of the resonant converter, voltage E_2 can be considered as a slowly varying DC voltage source. Under such circumstances current I_2 is really the response to some control parameter.

Thus, controlling the power flow amounts to controlling the average output current I_2. In the series resonant converter, rectifier CS_2 being a diode bridge, adjusting the power flow means adjusting current J in the LC resonant network.

In practice, the assemblage consisting of voltage inverter CS_1 and the LC series resonant network can be likened to a current source. Of course, this current source is not perfect and its amplitude depends on the operating point. It should be noted that in the special case of the discontinuous mode (Fig. 6.21: $f_s <$ $f_0/2$), it is the average value of full-wave rectified current J that is constant, but the peak current depends on the operating point.

Since the LC resonant network behaves as a selective filter, a simplified study taking only the fundamental terms into account should give results of sufficient accuracy, especially for frequencies f_s close to f_0, that is to say in the continuous conduction mode.

Under these assumptions, the equivalent current source is composed of a sinusoidal voltage source V_1 with an amplitude equal to the fundamental of voltage v_1 delivered by inverter CS_1, connected in series with an impedance $Z(\omega_s)$ equal to that of the LC resonant network at the frequency being considered (Fig. 6.34). Then, the different control techniques for adjusting the amplitude of current J can be more readily appreciated from Fig 6.34.

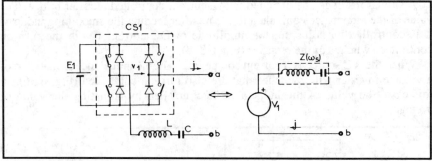

Fig. 6.34 Series resonant converter (the series combination of inverter CS$_1$ with a series resonant network is modelled as a sinusoidal voltage source in series with an impedance Z(ω_s))

(a) Control via Z(ω_S)

The first of these techniques involves varying Z(ω_s) while holding V$_1$ constant. This can be achieved in two different ways:
1. by varying ω_s, that is to say by varying the frequency, and
2. by varying the parameters of the LC network.

To be viable, this variation of the parameters of the resonant network must be achieved by electronic means. These techniques are similar to those used in static VAr compensators and are reviewed below.

Implementation of a variable inductor:

This generally consists of an inductor in series with two back-to-back turn-on controlled inherent turn-off switches (thyristors). Control of the firing angle of the switches enables adjustment of the amplitude of current j. Thus the voltage source 'sees' the equivalent of a variable inductor (Fig. 6.35a).

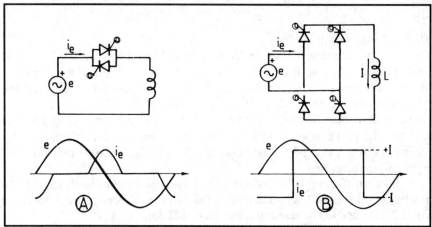

Fig. 6.35 Principle of implementation of a variable inductor

119

A similar result can be obtained with a current-controlled rectifier bridge. The value of the inductance equivalent to the rectifier feeding the smoothing inductor L is controlled by adjusting the amplitude of DC current I with the turn-on controlled switches of the rectifier (Fig. 6.35b).

When the AC source is a current source, the variable inductance is made up of a voltage inverter with switches that are also turn-on controlled (Fig. 6.36). In this case, the value of the inductance is controlled by adjusting the value of voltage V.

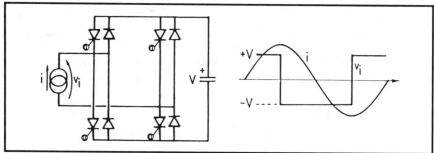

Fig. 6.36 Principle of implementation of a variable inductor: case of an AC current source

Implementation of a variable capacitor:

The different techniques used to make a variable capacitance are derived by duality from those used for variable inductance:

1. a capacitor connected in parallel with a turn-off controlled switch with bidirectional voltage and current capabilities,
2. a turn-off controlled current rectifier, and
3. a turn-off controlled voltage rectifier.

Only the last technique is suitable for circuits with an AC voltage source [87].

(b) Adjusting the amplitude of V_1

The second technique for controlling current J involves varying V_1 while holding $Z(\omega_s)$ constant (Fig. 6.34). However, in order to retain all the properties of the resonant converters, this control should not affect the commutation conditions of inverter CS_1. If all the switches of CS_1 have the same commutation mechanisms, adjustment of V_1 is difficult to achieve within inverter CS_1. Thus an auxiliary converter connected in front of CS_1 (controlling the value of E_1) or between CS_1 and the resonant network (AC regulator) must be used.

A more satisfactory solution is to use an inverter for CS_1 in which the two inverter legs have dual commutation mechanisms. Such a converter, shown on Fig. 6.37, is operated at the natural frequency [22,23].

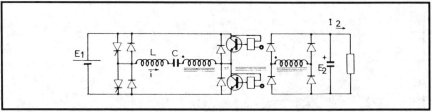

Fig. 6.37 Series resonant inverter. The two inverter legs use commutation mechanisms which are the dual of each other, which allows control of the energy flow at fixed frequency

(c) Phase shift control
The fixed frequency control techniques discussed above are not the only ones possible. Basically, controlling the power flow requires control of the current drawn by diode rectifier CS_2 rather than the current in the resonant network (though these two currents are identical to each other in the circuit of Fig 6.33a).

One solution is to consider not just one current source, as in Fig. 6.34, but two identical parallel connected sources, with a variable relative phase shift that can be controlled between 0 and π as indicated on the diagram of Fig. 6.38. This type of control is referred to as phase shift control (Appendix A5) [91~93].

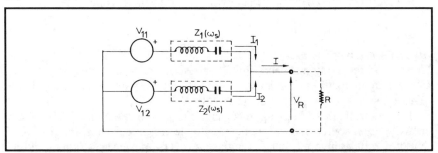

Fig. 6.38 Principle of phase-shift control in a series resonant inverter

The output characteristics $E_2(I_2)$ are ellipses whose semi-axes lengths are proportional to $\cos(\varphi/2)$ - φ being the phase shift between voltage sources V_1 and V_2. Experimental $E_2(I_2)$ characteristics are given in Fig. 6.39. A rapid increase of the output voltage for low values of I_2 can be seen, which is due to the discontinuous operation of rectifier CS_2.

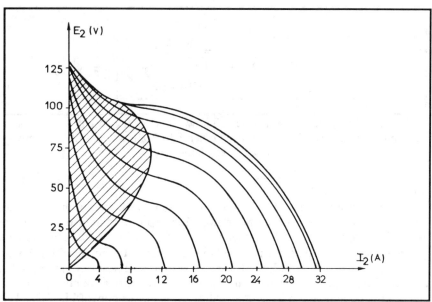

Fig. 6.39 Phase-shift control in a series resonant inverter:
experimental output characteristics corresponding to different values of the
phase-shift.

(d) Controlled rectification

Up to now, the study has been limited to a series resonant converter in which CS_2 is a diode rectifier. Replacing this diode rectifier with a controlled rectifier gives a new way of regulating the power flow.

This controlled rectifier (Fig. 6.40) is in the special situation of being fed by an AC current source and supplying a DC voltage source. Thus, its topology is that of a voltage inverter, but as long as no reverse power flow is required, the structure of rectifier CS_2 can be one of those indicated in Fig. 4.10.

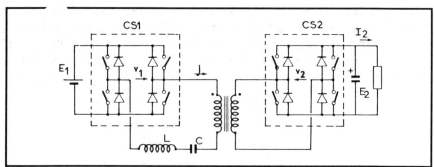

Fig. 6.40 Reversible series resonant converter:
the current rectification is controlled

The analysis of the reversible converter, already well developed by others [94~95], is given in Appendix A6 using an approach based very much on synthesis. Here, we will simply state that controlling the phase shift δ between voltages v_1 and v_2, that is to say between the control signals of inverters CS_1 and CS_2 effectively controls the amplitude of current I_2.

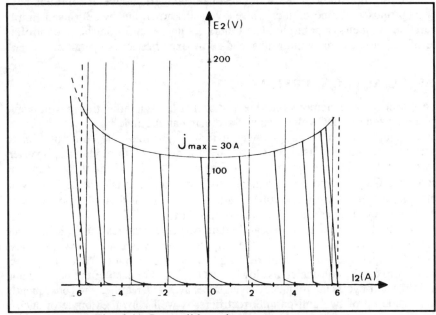

Fig. 6.41 Reversible series resonant converter:
experimental output characteristics corresponding to different values of the phase-shift between the control signals of CS_1 and CS_2

Experimental output characteristics of such a resonant converter are given in Fig. 6.41. The low power of this actual converter and the power control range (δ>π/2u, leading or lagging) leads to a significant overrating of the switches (Appendix A6) which results in non-negligible losses and explains the inclination of the experimental characteristics compared to the theoretical characteristics.

Also, when the output voltage is low, the residual voltage ripple causes unwanted conduction of the rectifier diodes and thus an uncontrolled increase of the output voltage. This phenomenon is to be compared, by duality, with that of the discontinuous conduction in voltage-fed rectifiers in which, under light load conditions, the current ripple causes an inherent and unwanted turn off of the switches and hence the output voltage increases.

6.4.3 *Conclusion*

By contrast with switch-mode converters in which the turn-on and turn-off commutations are controlled independently, resonant converters have a single controlled commutation. Consequently, resonant converters require specific power control techniques.

Limiting the study to the series resonant converter, several solutions have been proposed. Some of them, already well known and widely used, imply variable frequency operation. The others — phase-shift control or controlled rectification — are more innovative and allow fixed frequency operation.

6.5 DC/AC CONVERSION

This section will demonstrate the possibility of implementing a sine-wave generator fed by a DC voltage source and using an AC link.

This conversion technique, which aims to utilize soft commutation to perform the function usually realised by a a pulse-width-modulated inverter, corresponds to the diagram in Fig. 6.1. The converter CS_1 is an inverter and the converter CS_2 is a frequency changer with a structure that is to be defined. Note that phase-shift between the voltage and current is intrinsic in AC operation so the frequency changer must allow reverse power flow.

The frequency changer also bears certain particular features due to the intrinsic nature of the AC link converters: the output frequency is always very much less than the input frequency which also implies that the output (current or voltage) waveform can be considered as slowly varying DC. For each of the positive and negative half-cycles the frequency changer behaves as a rectifier and consequently it is made up of two anti-parallel rectifiers. A well known example of such a frequency changer is the cycloconverter.

A synthesis method similar in all respects to that used for DC/DC conversion (Section 6.2) consists in distinguishing two cases depending on the type of source S_2. All the techniques used in Sections 4.4.3 and 6.2 can be derived for DC/AC conversion. However, for all the reasons already stated in this book, we will focus on the structures using the series resonant converter.

Furthermore, the complexities inherent in the implementation and control of a direct frequency changer (four-segment switches, zero-crossing detection of the output) lead us to propose novel solutions using only voltage-source inverters which are well suited to operation at high frequencies. Thus, the electrical variables within these DC/AC resonant converters are DC voltages, possibly slowly varying, and AC currents [94~95].

6.5.1 The reversible series resonant converter

This converter uses a controlled rectifier in order to achieve reverse power transfer and is the basic constituent of all the circuits discussed in the following section. It is shown in Fig. 6.40 and studied in Appendix A6.

As soft commutation operation requires square-wave voltages for v_1 and v_2, only two control parameters are available. They may be, for example, the frequency f_s determined by the controller for inverter CS_1 and the phase-shift δ between the control waveforms for CS_1 and CS_2.

From a functional viewpoint, this circuit is equivalent to a pulse-width-modulated inverter leg equipped with an LC filter (Fig. 6.42), or to a current-reversible step-down chopper. However the reversible series resonant converter has a natural step-down/step-up capability and galvanic isolation and/or impedance matching can be achieved in very favourable conditions at the AC link level. In this sense, the series resonant converter behaves mostly as a reversible Flyback converter.

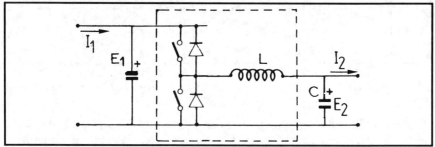

Fig. 6.42 The series resonant converter is, intrinsically, equivalent to an inverter leg equipped with an LC filter

6.5.2 Indirect conversion

To achieve DC/AC conversion with a reversible resonant converter, an intermediate 'DC' voltage V with the waveform of a full-wave rectified sinusoidal voltage is generated (Fig. 6.43). This intermediate 'DC' voltage V supplies an inverter of which the commutations are synchronized with the zero-crossings of V [94].

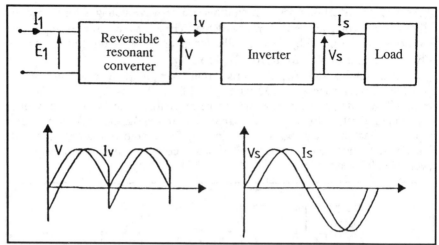

Fig. 6.43 DC/AC resonant conversion with folder-unfolder
(the series resonant converter is 'followed' by an inverter that commutates at each zero-crossing of voltage V)

In the light of duality, this first solution is reminiscent of those sine-wave generators which, when fed by an AC voltage source, make use of a rectifier, a smoothing inductance and a current-source inverter. If these sine-wave generators are fed by a DC voltage source, the rectifier is replaced with a voltage-reversible chopper, or else by an AC link converter such as that described in Section 4.4.3 (Figs 4.12 to 4.15).

In this type of conversion (Fig. 6.43), the resonant converter implements all of the power transfer control, namely the control of the amplitude and the frequency of the output voltage, while the voltage inverter simply performs the DC/AC conversion.

It should be noted here that the commutation mechanisms of the voltage inverter only depend on the inductive or capacitive nature of the load. Nevertheless, these commutations do not waste energy since they occur under zero voltage and at a frequency equal to the output voltage sine-wave. The design and implementation of this circuit is straightforward and does not raise any special problems.

The voltage ratings of the switches are on the one hand equal to the input DC voltage and on the other hand to the output peak voltage. The current rating of the switches in the inverter is determined by the load current while that of the switches in the resonant converter strongly depends on the operating range (Appendix A6). The resonant converter, having by design an operating frequency far greater than the output frequency, must also be designed for the peak instantaneous power it is liable to supply.

126

Apart from the ratings, the main design problem lies in the control and is due to the fact that each commutation of the inverter must cause a rapid reversal of the current delivered by the resonant converter. On the other hand, to achieve an accurate tracking of the sinusoidal reference, the regulator should include at least one integrator.

Experimental current and voltage waveforms obtained with an inductive load at 400Hz are given in Fig. 6.44. A slight distortion in the voltage waveform can be observed at the zero crossing.

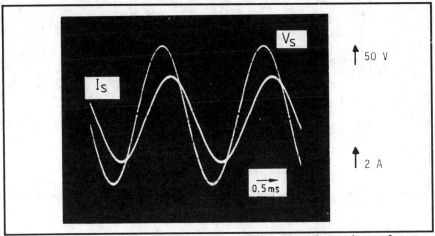

Fig. 6.44 DC/AC converter with folder-unfolder. Experimental waveforms

This principle of indirect resonant conversion is well suited for a single-phase generation. In contrast, it is apparently less attractive for polyphase generation, since it requires as many chains of converters as the number of phases.

6.5.3 Differential conversion

By virtue of the analogy developed in Section 6.5.1 between the reversible series resonant converter and the voltage inverter leg, a novel structure for a series resonant DC/single phase AC converter can be derived from the structure of a conventional inverter (Fig. 6.45). It consists of two reversible series resonant converters fed by the same voltage source E_1 with v_s (the voltage across the load terminals) being the difference between the voltages delivered by the resonant converters.

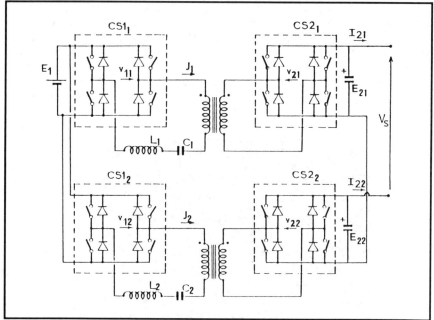

Fig. 6.45 Principle of DC/AC 'differential' resonant conversion: single phase case.

Denoting by V_s the amplitude of voltage v_s assumed to be sinusoidal, and by V_{DC} a DC voltage with an amplitude greater than $V_S/2$, the resonant converters are controlled so as to deliver the following output voltages:

$$E_{21} = V_{DC} + (V_S/2) \sin \omega_s t \qquad (6.26)$$

$$E_{22} = V_{DC} - (V_S/2) \sin \omega_s t \qquad (6.27)$$

which gives:

$$v_s = V_S \sin \omega_s t \qquad (6.28)$$

This differential conversion process can be extended easily to the generation of balanced polyphase voltages [91,94,95]. It requires as many resonant converters as the number of phases, plus one if the neutral needs to be brought out.

In the case of a three phase system (Fig. 6.46), if V_S is the amplitude of the line-to-line voltage, and V_{DC} a DC voltage greater than V_S, we have:

$$E_{21} = V_{DC} + V_S \sin \omega_s t \qquad\qquad (6.29)$$

$$E_{22} = V_{DC} + V_S \sin (\omega_s t - 2\pi/3) \qquad\qquad (6.30)$$

$$E_{23} = V_{DC} + V_S \sin (\omega_s t - 4\pi/3) \qquad\qquad (6.31)$$

$$E_{2N} = V_{DC} \qquad\qquad (6.32)$$

Though the differential conversion — like the indirect conversion — requires as many converters as the number of output phases, it does enable the removal of the output inverter. It thus solves radically the problem of the control, which in this case, must achieve accurate tracking of a sinusoidal voltage without rapid changes in the output current.

In a differential structure, the switches of rectifiers CS_2 of the resonant converters must withstand far greater voltages. The DC voltage V_{DC} must be larger than the amplitude of the sinusoidal voltage upon which it is superposed in order to allow good control of the output voltage and especially to avoid the discontinuous conduction described above (Section 6.4.2.4.). Thus, the switches must withstand voltages in excess of V_S in the case of a single-phase converter and in excess of $2V_S$ in the case of a polyphase converter. It is worth noting that the switches of a direct frequency changer generating the same three phase network voltage must withstand a voltage equal to $V_S \sqrt{3}$. From this viewpoint, the structure is not too disadvantageous.

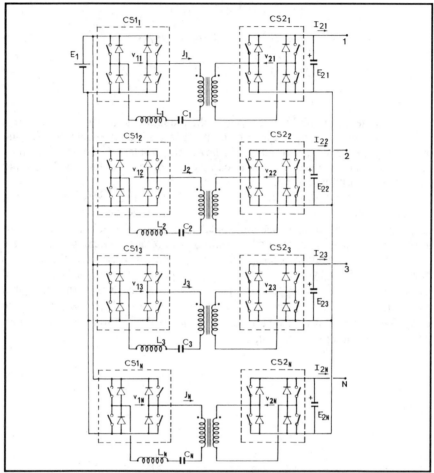

Fig. 6.46 Extension of the principle of DC/AC differential resonant conversion to three-phase plus neutral generation

As each resonant converter is always designed for the maximum power it has to handle, assuming a resistive load, the power absorbed by each phase is $V_s^2/2R$ whereas each resonant converter must be able to deliver $2V_s^2/R$. This calculation only involves perfect waveforms, but the current delivered by each phase is derived from the controlled rectification of a quasi-sinusoidal current. With full-wave rectification and making the most optimistic assumptions that the maximum current delivered by each phase is obtained with rectifier commutation occuring at the current zero-crossing and that the current is purely sinusoidal, the peak current in rectifiers CS_2 is $\pi V_s/2R$.

In the circuits of Figs 6.45 and 6.46, the operating frequencies of the various reversible resonant converters may be different. Assuming now that they are operated at the same frequency, the control of converters CS_1 can be in phase. Consequently, these CS_1 converters are operated in parallel so that a second circuit can be postulated with the switches all being paralleled within a single inverter CS_1 common to all the resonant converters, rather than paralleling several inverters (Fig. 6.47).

This DC/AC converter operates similarly to the former one but it has a very important advantage in the case of balanced polyphase loads. In effect, since the power P absorbed by such a load is constant, in the particular case of three phases we have at each instant:

$$\Sigma E_{2i} I_{2i} = P \qquad (i=1,2,3,N) \tag{6.33}$$

and thus:

$$(v_1 * \Sigma j_i)_{average} = P \quad (i=1,2,3,N) \tag{6.34}$$

and

$$E_1 I_1 = P \tag{6.35}$$

Under such circumstances, converter CS_1 need only be dimensioned for the actual power P that is transferred to the load, and the average current I_1 taken from the DC voltage source E_1 is constant.

Nevertheless, the maximum instantaneous power delivered by each phase circulates as far as the output of inverter CS_1 which means that currents flowing through each resonant network are not modified and the resonant networks do not benefit from operating with a polyphase balanced load.

Moreover, because of the three resonant networks and because all the converters in Fig. 6.47 are operated at the same frequency, it is very difficult to vary this frequency which must therefore be kept constant.

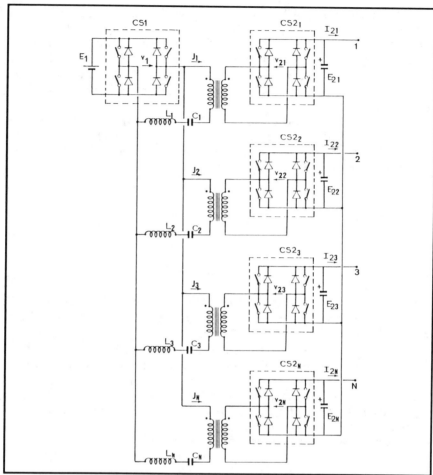

Fig. 6.47 DC/AC resonant converter with three-phase and neutral outputs: each output corresponds to one resonant network, one transformer and one controlled rectifier, all of them fed by a single inverter CS$_1$

Having thus lost a degree of freedom, one is driven to make a compromise regarding the regions of operation of the converters, taking into account their commutation limits, which inevitably leads to an increase in the resonant circuit currents and the current in inverter CS$_1$.

To overcome this problem, the scheme in Fig. 6.48 is proposed which can be extended to any number of output phases. A rectifier is associated with each of the output phases, and, if required, with the neutral. Each rectifier is then series-connected to the single resonant network through a single phase transformer that is indispensable in this case, and to inverter CS$_1$, which could equally well be connected to the resonant network through a single phase transformer.

132

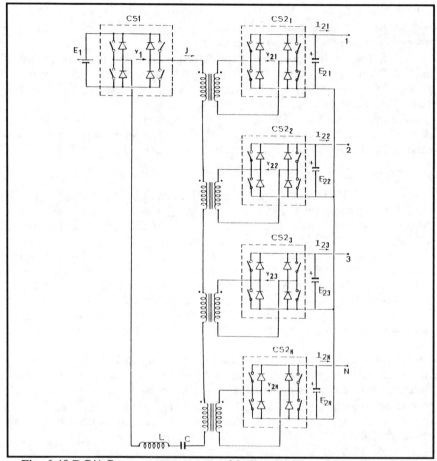

Fig. 6.48 DC/AC resonant converter with three-phase and neutral outputs: each output corresponds to one transformer and one controlled rectifier, and all the primary windings are connected in series to the resonant circuit and fed by inverter CS$_1$

The set of rectifiers CS$_2$s fed by current J flowing through the resonant network is in fact a frequency changer which is somewhat reminiscent of the well-known structure of the circulating current cross rectifier. By duality, we have here current-fed rectifiers that, instead of maintaining a DC current component, maintain a DC voltage component.

Still assuming a balanced polyphase load, the maximum instantaneous power delivered to each phase circulates only as far as the primary of the associated transformer and we have:

$$(j * \Sigma \ v_{2i})_{average} = P \quad (i=1,2,3,N) \qquad (6.36)$$

assuming the transformer ratios are one to one.

Converter CS_1 is still designed for actual power P that is transferred to the load, the average current taken from the DC source E_1 is still constant and only power P flows through the resonant network.

More generally, because the output frequency is always much less than the AC link frequency, converter CS_1 only supplies the instantaneous peak power delivered by all the phases together, whatever the output waveforms and whether the load is balanced or not.

Furthermore, with this DC/AC converter topology (Fig. 6.48) which only has one resonant circuit, it is again possible to vary the operating frequency and thus impose the overall performance of the system at each operating point.

6.5.4 Conclusion

Several DC/AC converters with an AC link have been proposed but basic considerations of power transfer show that the circuit of Fig 6.48 should have the best performance with respect to efficiency but also with regard to weight and size.

The function of these converters is equivalent to that of pulse-width-modulated inverters but they exhibit the extra feature of intrinsic filtering of the output voltage and of possible galvanic isolation through high frequency transformers. If the switching frequency of the converters is sufficiently high (a few tens of kilohertz, for instance), the output voltages can be modulated up to frequencies in the one kilohertz range with harmonic distortion around one per cent (Section 7.3.5).

Because of the reversibility of the various converters connected to the resonant network, the structure of Fig 6.48 is fully reversible and the overall power can flow either toward voltage source E_1 or from voltage source E_1. The next stage is to design resonant AC/DC converters, particularly rectifiers that absorb sinusoidal current from the mains with controllable phase and very low harmonic distortion.

6.6 GENERAL COMMENTS ON RESONANT CONVERSION

The topologies of non-reversible (Fig. 6.33a) and reversible (Fig. 6.40) DC/DC converters and of DC/AC converters (Fig. 6.48) have certain similarities which encourage a general approach, including by synthesis, to resonant conversion.

Assuming the switching frequency to be greater than the natural frequency and taking only the fundamental terms into account, the modelling of inverter CS_1 and of the resonant network is somewhat similar to that of an alternator. This leads us to the identification in each of these topologies of a single phase AC current supply generated by inverter CS_1 from a DC voltage source E_1 and filtered by the resonant network.

Power is taken from this resonant network by one or several converters which are voltage inverters operating in the current rectifier mode and which can be fed through single phase transformers.

If these rectifiers are not controlled (diode bridges), they can only supply DC loads, which are uni-directional with respect to both voltage and current. When they are controlled they are able to supply directly DC loads that are not voltage-reversible but are current-reversible. It has also been shown that AC loads with frequencies much less than the natural frequency can be supplied, and consequently DC loads with reverse voltage requirements, by using a particular inter-connection of the DC outputs (Fig. 6.48).

Thus, the most general topology for a resonant converter has the form of an AC network comprising a resonant circuit that exchanges energy between at least two electrical systems (Fig. 6.49). Each electrical system S_i is connected to one or more voltage inverters-current rectifiers Cv_i that may have associated transformers Tr_i. The connection of the DC terminals of these voltage inverter-current rectifiers depends on the reversibility of the electrical system to which they are connected.

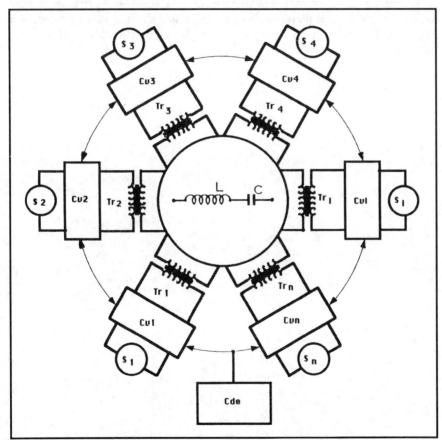

Fig. 6.49 Generalization of resonant conversion

Because of the high frequency operation, the resonant network can only store a limited amount of energy, which implies an overall control of the energy transfers between the different electrical systems. Thus, the control strategy for these resonant converters consists of transferring energy from one electrical system to the resonant network and then from the resonant network to another electrical system. These energy transfers are performed by choosing the commutation instants in such a way as to create the correct conditions for soft commutations, and, if possible to avoid storing too much energy in the resonant circuit.

Two schemes for the control of energy flow in resonant converters, based on this principle, have been patented with ANVAR [91,96].

Using these resonant conversion techniques it is possible to implement all the 'functions' of conventional converters. In particular, an AC/AC converter could be designed (Fig. 6.50) with two frequency changers like that of Fig. 6.48 interconnected via a resonant network, one being supplied by the fixed-frequency fixed-amplitude three phase mains, for example, and the other providing a polyphase (three phase in example) sinusoidal voltage supply with variable frequency and amplitude.

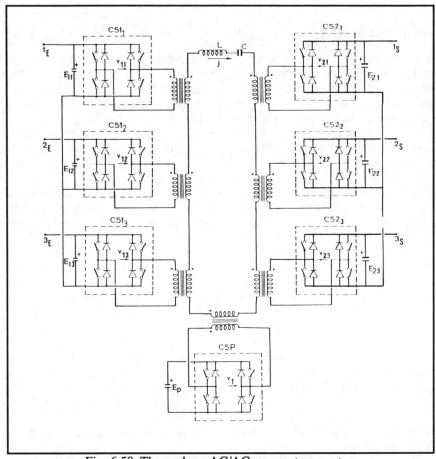

Fig. 6.50 Three-phase AC/AC resonant converter

In this AC/AC converter, both input and output frequencies must be less than the frequency of the AC link, but the ratio of input to output frequencies is not subject to any limitation.

A static VAr compensator can also be added to the resonant network (Section 6.4.2.(a)) as shown in Fig 6.50. Such compensation might be rendered necessary by the number of converters connected to the resonant network and using phase-shift control in order to enable operation in the soft commutation mode (Section 4.5).

6.7 CONCLUSION

In applying to AC link converters a very similar approach to the synthesis method for direct converters exposed in Section 1.6.1 (a), attention was first focused on non-reversible DC/DC conversion. A certain number of topologies have been catalogued, but one of them, the series resonant converter, stands out from the others because of its structure, features and performance.

Power control of this resonant converter can be achieved by means of several techniques, some requiring variable frequency operation, others fixed frequency operation. The static characteristics and the dynamic performance of resonant converters are strongly dependent on the control techniques used.

Among the control methods that require variable frequency operation, optimal control, which fixes the state plane trajectories, gives the best dynamic performance. In addition, optimal control accentuates the current source behaviour of the resonant converter.

Combination of this optimal control principle with phase-shift control of non-reversible resonant converters, or its extension to reversible resonant converters, makes fixed frequency operation possible again and also allows it to be considered for application to DC/AC or AC/AC resonant converters.

Like the non-reversible converter, these complex converters, albeit with extremely modular structures, conform to an overall general scheme for resonant link conversion. This approach to static conversion should lead to a number of original and high-performance converters, especially in the field of uninterruptible power supplies.

Within AC link converters, soft commutation and resonance are two techniques that allow the best utilization of today's power semiconductors. Extreme commutation frequencies can be reached with outstanding efficiency, although reliability and safety of operation are definitely the major advantages of these converters.

Conversely, the technical problem posed by the overall control of the energy flow within these converters is clearly a difficult one since it involves not only the generation and operation of an AC supply, but also its utilization within the scope of its imperfections.

7
DESIGN AND APPLICATIONS

7.1 INTRODUCTION

After the introduction and justification of several more or less well known and complicated conversion topologies based on series resonance, this chapter will focus on the design of the non-reversible series resonant converter. Following this, a brief description will be given concerning a collection of DC/DC and DC/AC converters developed under various industrial contracts by the Laboratoire d'Electrotechnique et d'Electronique Industrielle (L.E.E.I.), Toulouse, France.

Only the non-reversible series resonant converter will be considered, using three approaches that predict the converter operation with varying levels of accuracy:

1. The inverter and rectifier commutation phenomena are neglected and the current in the resonant network is assumed to be perfectly sinusoidal.

2.The commutation phenomena of both the inverter and rectifier are still neglected, but the real waveform of the current in the resonant network is considered.

3. A detailed analysis of the operation of the converter is made, especially taking into account the turn-off snubbers and the parasitic elements in the transformer.

The industrial applications of resonant converters described in this chapter concern:

• a prototype of a 40 kW (200 V-200 A) DC supply for a deflection coil in the LEP project at C.E.R.N. (Jeumont-Schneider),

• a high voltage (25 kW-140 kV) generator intended for the power supply of an X-ray lamp for radiography (Thomson-C.G.R.),

• a high voltage generator (5 kW-15 kV) supplying a CO_2 laser for the cutting of textile and sheet metal (LECTRA Systèmes),

• a 57 V-25 A battery charger (T.E.G.), and

• a low frequency three phase sinewave generator with 1 kVA capacity and very low harmonic distortion (T.E.G.).

7.2 DESIGN OF RESONANT CONVERTERS

For all the reasons already given in this book, we will focus on the series resonant converter operating above the resonant frequency.

7.2.1 Statement of the problem

As a general rule, the design of a static converter consists mainly of establishing the various stresses applied to its component elements. These stresses are deduced from the specifications that, in particular, define the output power and the conditions under which it should be delivered, the input voltage range, and sometimes, the operating frequency implicit in the choice of technology.

In the particular case of the series resonant converter (Fig. 7.1), the voltage ratings of the switches in the inverter (CS_1) and the rectifier (CS_2) are fully determined by voltages E_1 and E_2 given in the specification. Thus, the design involves choosing capacitor C and inductor L of the resonant network, the capacitors of the turn-off snubbers C_1 and C_2, the turns ratio k and the current rating of switches in CS_1 and CS_2.

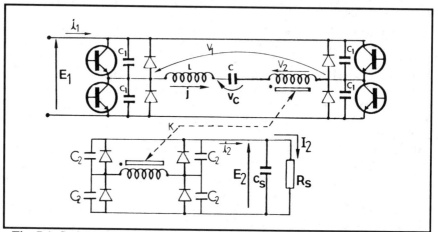

Fig. 7.1 Series resonant converter (the inverter and the rectifier are equipped with snubber capacitors)

Since the stresses applied to the switches depend not only on the specification but also on the values of C, L, C_1, C_2 and k, a design method leading to minimal stress on the switches for a given specifications is necessary. Such a design must:
• minimize the current in the resonant network and in the switches in order to minimize the losses and to obtain the highest efficiency, and
• minimize the kVA product of the components to reduce their size and weight. Thus, the overvoltage across the capacitor or the inductor is an important parameter.

To achieve such a design, the switches are assumed to be perfect, the input and output voltages E_1 and E_2 to be perfectly smooth, and the inductors and the capacitors to be lossless. The leakage inductance of the transformer is included in inductor L of the resonant network and the magnetizing current is neglected. Under these assumptions, the results from the steady state analysis of the series resonant converter can be used (Appendix A2 and A3).

Finally, the design must be worst case, that is to say with the lowest input voltage, assumed constant and equal to E_1, and for maximum (nominal) power P_n. This power corresponds to an output voltage of E_{2n}, an output current equal to I_{2n} and a switching frequency of f_{sn}.

7.2.2 Preliminary remark

Though capacitors C_1 and C_2 have a non-negligible influence on the output characteristics (Appendix A3), their primary function as snubber capacitors must always take precedence.

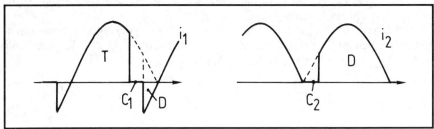

Fig. 7.2 Ideal waveforms of the current at the input and output of the series resonant converter

Assuming sinusoidal current in the resonant network, the input and output currents i_1 and i_2 are shown in Fig 7.2. The zero plateau in current i_1 (respectively i_2) corresponds to the commutation of converter CS_1 (respectively CS_2). The duration of this commutation increases as the snubber capacitor is increased, assuming everything else remains identical. Capacitors C_1 and C_2 play a part very similar to that of the commutating inductances in a voltage rectifier and must be minimized to reduce the amplitude of the current in the resonant network.

7.2.3 Analysis based on the fundamental component

In all this section, capacitors C_1 and C_2 will be neglected and the amplitude of the fundamental of an instantaneous quantity x will be denoted by X.

Assuming the current in the resonant network to be perfectly sinusoidal (due to the filtering effect of the LC resonant network), only the fundamental components of the voltages v_1 and v_2 are involved in the power flow.

Voltages v_1 and v_2 are square-waves with respective amplitudes of E_1 and E_2/k. The commutations of the diode rectifier CS_2 occur at the zero crossings of current J which causes the fundamental components of voltage v_2 and current J to be in phase. The circuit composed of the transformer, the rectifier, the filter and the load thus behaves like a resistance. V_2 is related to the output voltage by:

$$V_2 = 4E_2/k\pi \qquad (7.1)$$

and J to the output current by:

$$J = \pi \, k I_2 / 2 \qquad (7.2)$$

From a vector diagram for the AC circuit, the fixed frequency characteristic $V_2(J)$ can be derived:

$$V_2{}^2 + (\, J\sqrt{L/C}\,\,)^2 * (u - 1/u)^2 = V_1{}^2 \qquad (7.3)$$

where: $\qquad\qquad\qquad u = f_s / f_0 \qquad\qquad\qquad (7.4)$

This characteristic, showing the influence of the variation of voltage V_2 on current J at a given frequency, is an ellipse.

Note that for a given frequency, the power delivered by this converter is naturally limited and that as voltage V_2 varies from 0 to V_1, the maximum power P_{max} able to be delivered to the load is given by:

$$P_{max} = \frac{V_1^2 \, \sqrt{C/L}}{4\,(u\text{-}1/u)} \qquad (7.5)$$

At this full power operating condition, the power factor of inverter CS_1, considered here as a sinewave generator, is V_2/V_1, that is to say E_2/kE_1, and is equal to $1/\sqrt{2}$.

(a) Full-range operation
If operation from open circuit to short circuit is required (welding applications, for instance), the nominal point is naturally placed at this maximum power point. The turns ratio is then fixed:

$$k = (E_{2n}/E_1)\sqrt{2} \qquad (7.6)$$

Current J reaches its peak in short-circuit conditions and is:

$$J_{max} = \frac{V_1 \, \sqrt{C/L}}{(u_n\text{-}1/u_n)} \qquad (7.7)$$

or else:

$$J_{max} = \pi \, P_n/E_1 \qquad (7.8)$$

Current J_{max} is fully determined by the power P_n and voltage E_1, both defined by the specification. It does not depend on the value of L and C, but does determine the current ratings of the switches in converters CS_1 and CS_2.

At this stage, f_{sn}, the most suitable operating frequency for a given technology, can be defined. From the specification, we can also determine:
• current I_n that develops power P_n with an input voltage equal to E_1:

$$I_n = P_n/E_1 \qquad (7.9)$$

• capacitor C_n that provides a reactive power P_n in the form of a sinewave voltage with RMS value E_1 and angular frequency ω_{sn}:

$$C_n = P_n/E_1{}^2\, \omega_{sn} \qquad (7.10)$$

• inductor L_n that absorbs a reactive power P_n in the form of a sinewave voltage with RMS value E_1 and angular frequency ω_{sn}:

$$L_n = E_1{}^2/P_n\, \omega_{sn} \qquad (7.11)$$

Current J_{max}, which, from (8) and (9) is given by:

$$J_{max} = \pi\, I_n \qquad (7.12)$$

in effect defines the impedance of the resonant network at the operating frequency but there is no way it can determine the values of L and C directly from the specification.

The resonant converter designer is thus led to use another criterion to obtain a second relation between L and C. With this aim in view, it is important to remember that a specific feature of the resonant converter is to achieve power regulation by adjusting the switching frequency.

As a variation Δu in the frequency causes a variation ΔP in the power, the ratio $\Delta P/\Delta u$, which sets the range of the power control, is an important characteristic for the operation of the resonant converter. Generally speaking, a high value is desired so as to reduce the frequency variation.

However, for the design, it is preferable to replace this parameter by a quantity to which it is related but that is more conveniently used, namely S, the normalized peak voltage across the capacitor. This value of S varies with the power control gain and determines the voltage ratings for both the capacitor and the inductor in the resonant network. Let:

$$S = V_{Cmax}\, /E_1 \qquad (7.13)$$

Thus, knowing S leads directly to V_{Cmax} on one hand, and to C and L on the other, since:

$$C = C_n\, \pi/S \qquad (7.14)$$

$$L = L_n\, (4 + \pi S)/\pi^2 \qquad (7.15)$$

Considering the frequencies, the peak voltage across capacitor C, given by:

$$V_{Cmax} = \frac{4\, E_1}{\pi\left(u_n^2 - 1\right)} \qquad (7.16)$$

becomes even greater — and thus the power control gain even better — the closer the operating frequency f_{sn} is chosen to the natural frequency f_0.

143

(b) Limited load current
When the load current is limited, either by the load resistance or by the control, the design proceeds somewhat differently.

Since the series resonant converter naturally steps down the voltage, the turns ratio is chosen such that:

$$k = E_{2M}/E_1 \qquad (7.17)$$

where E_{2M} is the maximum value of voltage E_2. The maximum current in the resonant network is given by:

$$J_{max} = \pi \, k \, I_{2M}/2 \qquad (7.18)$$

where I_{2M} is the maximum value of current I_2. E_{2M} and I_{2M} can a priori take values that differ from E_{2n} and I_{2n}.

The voltage and current in the switches are then fully determined and the designer can choose the most suitable operating frequency f_{sn}.

The resonant frequency, which is equal to the frequency f_{sn} when E_2 equals E_{2M}, is also predetermined. Nevertheless, it does not completely determine L and C, and the designer must also take some extra criterion into account such as the peak voltage across the capacitor.

The determination of the operating point that yields the maximum peak voltage across capacitor C is a function of the area over which the converter load may vary. Knowing the normalized peak voltage S allows the value of C to be calculated — and hence L from the equation for the resonant frequency. In the particular case when the converter behaves as a voltage source with current limiting, we have:

$$J_{max} = I_n \, \pi/2 \qquad (7.19)$$

$$C = C_n \pi/2 \, S \qquad (7.20)$$

$$L = L_n 2S/\pi \qquad (7.21)$$

Assuming that everything else remains the same, comparing the two expressions for J_{max} (7.12) and (7.19) shows that the natural short-circuit capacity of the converter results in a doubling of the current in the various elements.

7.2.4 State plane analysis

Normalized state plane analysis of the resonant converter, with capacitive snubbers C_1 and C_2 neglected (Appendix A2), leads to an analytic equation for the output characteristic:

$$q^2 \sin^2\left(\frac{\pi}{2u}\right) + \left(1 + Y_{2avg}\frac{\pi}{2u}\right)^2 \cos^2\left(\frac{\pi}{2u}\right) = 1 \qquad (7.22)$$

This equation is also that of an ellipse. On this characteristic, the peak power is delivered to the load when:

$$Y_{2avgM} = \frac{3u}{2\pi}\left(-1+\sqrt{\frac{8}{9}\tan^2\left(\frac{\pi}{2u}\right)+1}\right)$$

(7.23)

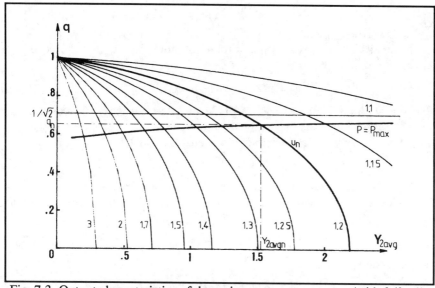

Fig. 7.3 Output characteristics of the series resonant converter (with $f_s/f_0>1$) showing locus of the maximum power point

The locus of the maximum power point in the $q(Y_{2avg})$ plane is plotted in Fig 7.3. The normalized output voltage q lies between 0.58 and 0.7 and is thus always less than the value obtained based on the fundamental component only ($1/\sqrt{2}$). Furthermore, the ratio of the value of the current in the resonant network at full power to that in the short circuit mode lies between 0.7 and 0.66 and is thus always less than the value given by calculations based on the fundamental component($1/\sqrt{2}$). This is easily explained by considering the form factor of the current in the resonant network.

Assuming input voltage E_1 to be constant, the normalized peak voltage determines the voltage the capacitor must withstand. This normalized peak voltage, denoted by x_m, is equal to S based on the usual assumption regarding the fundamental. It is related to the normalized output voltage q and to the normalized frequency u by the following relation:

$$q^2\sin^2\left(\frac{\pi}{2u}\right)+\left(1+x_m\right)^2\cos^2\left(\frac{\pi}{2u}\right) = 1$$

(7.24)

145

In the $q(x_m)$ plane, this equation is that of an ellipse centred at point (-1,0) and passing through the fixed point (0,1). For a given normalized frequency u, the maximum normalized peak voltage reaches its peak in the short circuit mode and equals:

$$x_M = (1 - \cos(\pi/2u))/\cos(\pi/2u) \qquad (7.25)$$

(a) Full-range operation
If the load can vary from open-circuit to short circuit, the choice of x_M effectively fixes u_n from:

$$u_n = \pi/2 \arccos(1/(1+x_M)) \qquad (7.26)$$

and thus defines in the $q(Y_{2avg})$ plane the characteristic that contains the nominal point corresponding to the maximum power delivered to the load (Fig. 7.3). Then, the voltage conversion ratio is fully determined:

$$k = E_{2n}/E_1 q_n \qquad (7.27)$$

Also, from the state plane analysis, the maximum current in the resonant network can be obtained for the short-circuit case and at the lowest frequency (u_n) from:

$$j_{max} = E_1\sqrt{C/L}\tan(\pi/2u_n) \qquad (7.28)$$

which becomes:

$$j_{max} = I_n \tan(\pi/2u_n)/q_n Y_{2avgn} \qquad (7.29)$$

in which Y_{2avgn} is deduced from Fig. 7.3 or from relation (7.23).

This fully specifies the switches and the frequency f_{sn} can now be determined by the designer in relation to the technology. The values of L and C are derived from the following two relations:

$$\sqrt{LC} = u_n/2\pi f_{sn} \qquad (7.30)$$

$$\sqrt{L/C} = E_1 Y_{2avgn}/kI_{2n} \qquad (7.31)$$

or as functions of C_n and L_n defined above:

$$C = C_n u_n/q_n Y_{2avgn} \qquad (7.32)$$

$$L = L_n u_n q_n Y_{2avgn} \qquad (7.33)$$

(b) Limited load current
If the load current is limited by the load resistance or by the control, the design process is very similar to that developed by considering the fundamental only.

The voltage conversion ratio is still defined by equation (7.17). The normalized peak voltage x_m, related to the output current by the following equation:

$$x_m = (\pi/2u)(kI_2/E_1\sqrt{C/L})\qquad(7.34)$$

reaches its maximum at the lowest frequency and the highest current I_2. Equation (7.34) gives the ratio $\sqrt{L/C}$ and we get, in the case of a voltage source with current limiting ($u_n=1$):

$$\sqrt{L/C} = (2x_ME_1)/(\pi kI_{2n})\qquad(7.35)$$

As long as the switching frequency equals the natural frequency, the current is perfectly sinusoidal and equation (7.18) remains true. However, because of its form factor, current J in the resonant network is maximum in the short-circuit case when I_2 is maximum also. The behaviour of the current j_{max} at short-circuit as a function of the current I_2 is shown in Fig 7.4. Approximately:

$$j_{max} = 1.7\,k\,I_2\qquad(7.36)$$

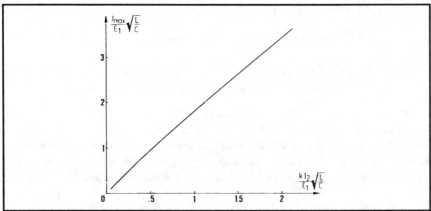

Fig. 7.4 Short-circuit operation of the series resonant converter: peak current in the resonant circuit j_{max} versus output current I_2

The designer can then choose the most suitable switches and determine the nominal frequency of operation , which is in this case the natural frequency, i.e. determine the product $\sqrt{LC}$ Equations (7.20) and (7.21) still apply if S is replaced by x_M, and they give the values of L and C.

7.2.5 Influence of the snubber capacitors

(a) Converter operation

The capacitive snubbers of inverter CS_1 introduce a commutation limit in inverter CS_1 (Appendix A3) that makes no load operation impossible, but they only modify the shape of the characteristics in the feasible operating area a little (Fig. 7.5) and have no influence on the maximum voltage and current stresses on the various components in the converter. However, this commutation limit is all the more severe if the capacitors have a large value of capacitance C_1.

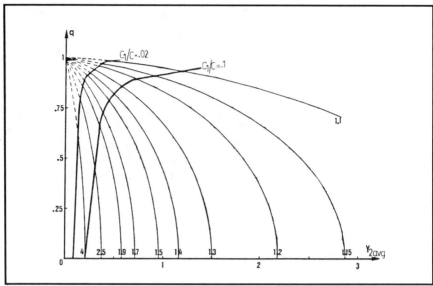

Fig. 7.5 Output characteristics of the series resonant converter (with $f_s > f_0$): influence of the snubber capacitors of the inverter

The capacitive snubbers of rectifier CS_2, or the equivalent parasitic capacitance connected on the secondary side of a high-ratio step-up transformer, allow no-load operation when the value C_2 is greater than that of the capacitors in inverter CS_1 referred to the secondary side (Appendix A3). The static characteristics and the commutation limits are then of the form indicated in Fig. 7.6.

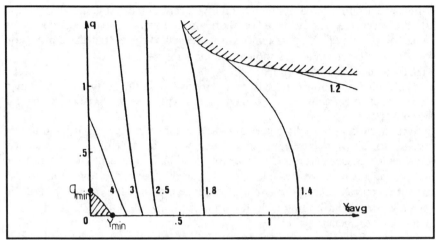

Fig. 7.6 Output characteristics of the series resonant converter (with $f_s/f_o>1$): some no load operating conditions are attainable when the snubbers in the rectifier are greater than those in the inverter

The no-load voltage, which is no longer directly fixed by the input voltage E_1, is controlled by the frequency. However, in the vicinity of the origin, there remains a zone in which the converter cannot operate. This zone is characterized by the minimum short-circuit current:

$$Y_{min} = \frac{2}{\Pi} \sqrt{\frac{C_1}{C+C_1}}$$

(7.37)

which only depends on the ratio C_1/C and on the minimum no-load output voltage:

$$q_{min} = C_1/k^2C_2$$

(7.38)

Besides modifying the no-load operation, capacitors C_2 make the characteristics more vertical and the series resonant converter really behaves as a current source over a certain frequency range.

A very qualitative explanation of this phenomenon is as follows: the transformer-rectifier (including capacitors C_2)-load behaves like a resistance-capacitance network. Effectively, therefore inverter CS_1 supplies a circuit whose resonant frequency is greater than that of the LC network, and other things being equal, the average current in the load is in this case greater than that obtained without the capacitors C_2.

It should be noted that opposite conclusions may be drawn if the no-load operation is obtained by means of the magnetizing inductance of the transformer.

(b) Converter design
The choice of the switch technology is made with regard to:
• the voltage rating, which is fixed by the specification,

149

• the current rating fixed by the maximum current in the resonant network. The latter always reaches its peak in short-circuit operation (independently of capacitors C_1 and C_2) and is almost proportional to the output current I_2 (Fig. 7.4). It is fully determined as soon as the turns ratio k is chosen.
• the frequency response required.

The choice of the turns ratio is an important step in the design, and as shown above, it depends mainly on the behaviour and type of characteristics desired for the converter.

Within the scope of characteristics such as those in Fig. 7.6, it is no longer possible to follow a complete output characteristic because of the commutation limit, which causes inverter CS_1 to cease functionning, or because of excessive over-voltages which are prone to appearing across the load at low values of the current I_2 (current source behaviour). Consequently, only the case of the voltage source with current limiting is dealt with in the remainder of this section, which is in fact a very general case!

In view of the preliminary remark (7.2.2), the rectangular load area of this source is entirely within the feasible operating region if:

$$k \geq E_{2n}/E_1 \qquad (7.39)$$

Furthermore, in order to reduce the amplitude of the current in the resonant network and in the switches, the maximum power point must obviously be chosen as close as possible to the commutation boundary of the inverter in order to minimize the conduction of the inverter diodes. For this reason, we choose:

$$k = E_{2n}/E_1 \qquad (7.40)$$

The maximum current in the resonant network can be deduced from Fig. 7.4, which allows full characterization of the switches, and also of the transformer, the snubber capacitors and gives the nominal frequency of operation.

In order to minimize Y_{min} and q_{min} (Fig. 7.6), from equations (7.37) and (7.38) the capacitors C_1 must be designed to be as small as possible. By contrast, capacitors C_2 can be chosen by the designer as a function of the commutation, of the minimum no-load output voltage (q_{min}), or moreover, they may be determined by the transformer technology.

The normalized peak voltage across capacitor C is given by (Appendix A3):

$$x_m = (\pi/2u)Y_{avg} + a_2q \qquad (7.41)$$

It reaches its maximum at the maximum power point (q=1). This maximum peak voltage can be expressed as a function of the real quantities:

$$x_M = \{(kI_{2n}/4f_{sn}E_1) + k^2C_2\}/C \qquad (7.42)$$

This latter equation leads to an expression for C as a function of C_n (7.10):

$$C = C_n\pi/2x_M + k^2C_2/x_M \qquad (7.43)$$

150

At this point, it is possible to plot the set of output characteristics since these only depend on the normalized quantities C_1/C and $k^2 C_2/C$ and it only remains to determine the value of the inductor L. On this set of curves, the point that represents the maximum power has an ordinate q_n equal to 1, an abscissa Y_{2avgn} and belongs to the characteristic with parameter u_n. Y_{2avgn} and u_n must fulfill the following relation:

$$Y_{2avgn} = (C_n/C)\, u_n \qquad (7.44)$$

The value of L is then very easily deduced.

7.2.6 Conclusion

In this section devoted to the design, it has been shown that a resonant converter can be fully designed from its specification with the aid of a few simplifying assumptions, notably by neglecting the snubber capacitors in the inverter and the rectifier.

Because of the influence of these capacitors on the overall operation of the converter (Appendix A3), the resulting design often constitutes a first approach, but it can, if required, be the start of an iterative process for specifying the different components in the converter.

However, by using the results of a detailed analysis of the series resonant converter, taking account of capacitors C_1 and C_2 and, if necessary, the parasitic elements of the transformer, a simple and accurate method of design for all the components of the converter, including capacitors C_1 and C_2, has been proposed.

7.3 APPLICATIONS

7.3.1 Power supply for an electromagnet

In 1982, C.E.R.N. invited bids from European power electronics companies concerning the construction of a high precision 40 kW (200 V-200 A) DC supply for the beam-steering electromagnets in its accelerators (LEP project) [97].

This supply, connected directly to the 380 V/50 Hz mains had to have small size, low losses and the best possible dynamic performance in order to maintain the beams within the accelerators which are very sensitive to perturbations. At the same time, the electromagnetic interference had to be kept to a minimum.

In 1984, Jeumont-Schneider and the L.E.E.I. took up the challenge and decided to construct a prototype in collaboration with the C.E.R.N. [98,99].

This power supply is a resonant converter, of the type shown in Fig. 7.1. The design lead to the following values:

$$C_1 = 88 \text{ nF} \tag{7.45}$$

$$C_2 = 1760 \text{ nF} \tag{7.46}$$

$$C = 2.5 \text{ }\mu\text{F} \tag{7.47}$$

$$L = 55 \text{ }\mu\text{H} \tag{7.48}$$

and to a voltage step-down transformer with a 2:1 turns ratio.

With operating frequency lying between 20 kHz (on load) and 66 kHz in the required power range, the power transistor is the most suitable device. Operation above the natural frequency — i.e. in the dual thyristor mode — was selected and the output characteristics are precisely those of Fig. 7.6.

Each switch of the inverter is implemented with 7 (6+1) MJ16022 (Motorola) transistors chosen for their high gain and very fast commutation, connected in antiparallel with a fast recovery BYT 67800 diode (Thomson).

The transformer, rated at 50 kVA , has a ferrite magnetic core and insulated multiple conductor windings.

In order to make the converter fully symmetrical, the resonant network is composed of two identical resonant circuits (25 μH and 5 μF) connected on each side of the transformer. The leakage inductance of the transformer (5 μH) adds to the inductance of the two resonant networks. The inductors use a technology similar to that of the transformer and all capacitors are polypropylene capacitors.

The operation of the converter at nominal input voltage, at full power (40 kW) and at reduced power (900 W), are illustrated in Figs 7.7 and 7.8 respectively where V_{ce} is the voltage across one switch and J the current flowing on the primary side of the transformer.

At C.E.R.N., the converter was connected to a conventional electromagnet and incorporated into a high precision regulation loop. Though the tests were passed successfully, another less sophisticated resonant solution using GTOs and giving far lower performance, especially as far as frequency response is concerned, was chosen by C.E.R.N. for the supply of the resistive windings in the LEP project. Nevertheless, the converter described above with a 10 V/2000 A output stage and phase shift power control (Section 6.4.2.3) was selected to supply the superconducting windings.

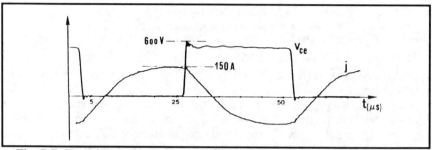

Fig. 7.7 Experimental waveforms of the output voltage and current of the
inverter in a series resonant converter (P = 40 kW)

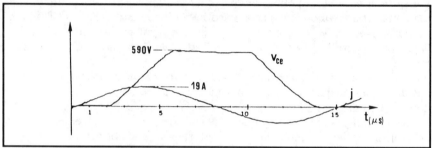

Fig. 7.8 Experimental waveforms of the output voltage and current of the
inverter in a series resonant converter (P = 900 W)

7.3.2 Power supply of an X-ray lamp

This high voltage power supply (25 kW-140 kV) for the X-ray lamp of a
scanner was studied in collaboration with Thomson-C.G.R. [66].

Scanning is a medical technique that enables indirect reconstitution of a cross-
section of the body from elementary measurements of radiological densities.

These measurements depend on X-ray absorption by the body and a thorough
knowledge of the laws of absorption, which requires strictly mono-chromatic X-
rays. This condition calls for a very stable DC voltage across the X-ray lamp to
avoid noise or defects in the image.

The voltage used is around 130 kV and the lamp requires a power of between
10 and 15 kW for the period of data acquisition, which is typically a few
seconds.

Previously, the high voltage generators used low frequency transformers
(50~60 Hz) to boost the mains voltage to the desired level. The control and high
precision regulation of the voltage were then achieved by two series-connected
tetrode valves on each side of the load, one on the positive side and the other on
the negative side. Though these techniques gave the desired performance, the
equipment was too bulky and expensive.

The requirements for these supplies are as follows:
• Supply: Three phase 380 V, 50 Hz mains,

- Output: 145 kV voltage with centre tap,
  220 mA current for a maximum duration of 18 s,
  power 25 kW,
- Stability: ±30 V for frequencies up to 150 Hz.

Reliability should be maximized, the efficiency as high as possible and the response time as short as possible.

Logical reasoning, based on analysis of the specification showed that the series resonant converter is the most suitable configuration for supplying this X-ray lamp. It should be noted here that the X-ray lamp, which behaves naturally as a current source, is perfectly stable when directly connected to the filter capacitor at the output of the resonant converter (voltage source).

Insofar as it renders the filter capacitor indispensable, the very high output voltage can be achieved either by obtaining the voltage step-up entirely within the transformer, which is then followed by a diode rectifier, or by obtaining a proportion of the voltage ratio within the transformer and following it by a capacitive multiplier [70]. This second solution was chosen for two major reasons:
- the diodes of the multiplier operate at lower voltage than those of the rectifier, and the voltages across the diodes are naturally balanced, and
- the transformer has a lower turns ratio and is consequently more easily built.

The resulting converter is symmetrical and composed of two series-connected voltage multipliers, each of them fed by a step-up transformer with 40:1 turns ratio with the primary windings connected in parallel (Fig. 7.9). The voltage multipliers are of the full-wave type to reduce voltage ripple. If this ripple is negligible, the voltages across the capacitors are constant and equal to a fraction of the output voltage that depends on the number of stages in the multiplier.

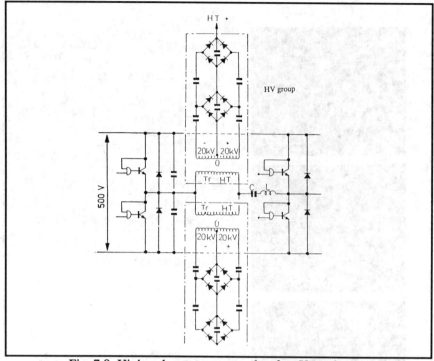

Fig. 7.9 High-voltage power supply of an X-ray lamp:
structure of the converter and HV sections

When fed by an AC current source, this transformer-multiplier set behaves exactly like a transformer which implements all the voltage step-up connected to a diode rectifier supplying a theoretically infinite filter capacitor.

The frequency must be sufficiently high to enable inaudible operation and reduce the size of the energy storage elements, especially the capacitors of the multiplier thus obtaining a very fast response. All these reasons indicate that a minimum frequency of operation of around 20 kHz is a reasonable compromise for such a power rating.

This power supply was engineered by Thomson-CGR and operates in the 18~50 kHz frequency range. Each switch is made up of 11 (10+1) MJ 16012 transistors (Motorola) and is capable of switching 120 A. Oscillograms of the voltage across a switch and of the current in the resonant network are given in Fig. 7.10 for several conditions of operation.

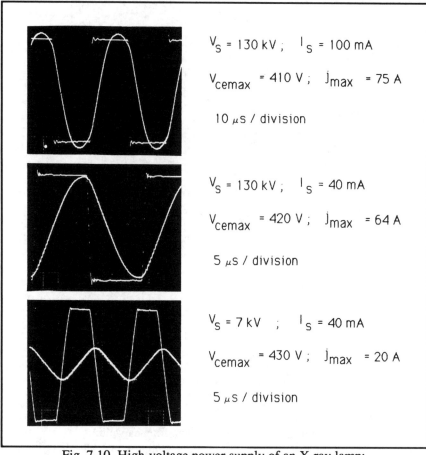

$V_S = 130 \, kV \, ; \quad I_S = 100 \, mA$

$V_{cemax} = 410 \, V \, ; \quad j_{max} = 75 \, A$

$10 \, \mu s \, / \, division$

$V_S = 130 \, kV \, ; \quad I_S = 40 \, mA$

$V_{cemax} = 420 \, V \, ; \quad j_{max} = 64 \, A$

$5 \, \mu s \, / \, division$

$V_S = 7 \, kV \quad ; \quad I_S = 40 \, mA$

$V_{cemax} = 430 \, V \, ; \quad j_{max} = 20 \, A$

$5 \, \mu s \, / \, division$

Fig. 7.10 High-voltage power supply of an X-ray lamp:
experimental current and voltage waveforms at the output of the inverter

The high voltage stage comprises the high voltage transformer and the multiplier stages which are oil-insulated. The high voltage capacitors have been specially designed to reduce their size.

The nominal power of 26 kW (130 kV/200 mA) was obtained with a 430 V input voltage and 112 A peak current in the resonant network.

In this application, the resonant technique yielded significant size and weight reductions (2.1 m^3 -> 0.9 m^3, 1230 kg -> 380 kg), an excellent efficiency (96% between the inverter input and the lamp), the required voltage stability owing to the high frequency operation, with high reliability and versatile control.

7.3.3 Power supply for a CO_2 laser

The laser effect is obtained by stimulating a gaseous mixture by an electric discharge within a cavity. To this extent, it behaves as a discharge lamp and operates in the region where the dynamic resistance is negative.

The CO_2 laser used by LECTRA Systèmes for the cutting of textiles or of sheet metal is composed of four electrically isolated guns (Fig. 7.11). These guns are optically in series in order to sum the luminous powers. The laser beam is focused on the working point by means of a reflecting mirror.

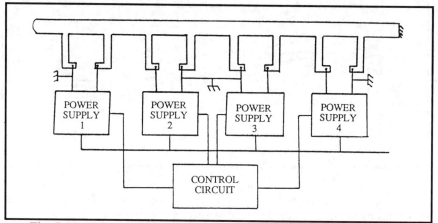

Fig. 7.11 Power supply arrangement for a CO_2 laser with four 'guns'

Each gun has its own power supply delivering a nominal power of 1 kW with a voltage around 15 kV, but the current can vary from 5 mA to 70 mA. This ionizing current is controlled by each power supply in relation to a common reference generated by the control circuit (Fig. 7.11).

The existing power supplies used a technology very similar to that used for X-ray lamps, namely a 50 Hz transformer, a diode rectifier, and a tetrode-based series regulator controlling the ionizing current. In addition, a load resistance was connected in series with the laser to limit the discharge current at switch-on for the period when the regulation is inoperative. Thus, these power supplies were heavy and bulky, efficiency was poor and reliability was greatly reduced because of the tetrode.

We first tried a resonant converter similar in all respects to that of Section 7.3.2, with a step-up transformer and a capacitive multiplier, and were confronted with two difficulties:

• The limiting series resistance had to be retained because the capacitive multiplier gives the resonant converter a voltage source characteristic and the laser presents a negative dynamic resistance.

• The laser characteristics (Fig. 7.12) vary over a very wide range with variations of pressure in the cavity and since the capacitive multiplier behaves dynamically as a voltage source, the ionizing current depends strongly on the regulation of the pressure in the cavity. The ionizing current variations reached 100% and could not be easily compensated for by the control since it was occuring in the load itself, downstream of he capacitive multiplier.

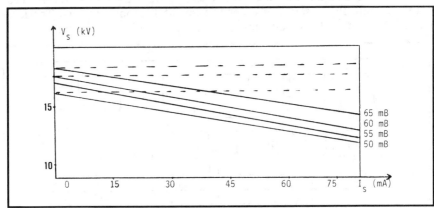

Fig. 7.12 Static characteristic of the CO_2 laser

Consequently, the capacitive multiplier was replaced with a single 150:1 step-up transformer and a diode rectifier directly supplying the laser without any filtering element (Fig. 7.13). It was then possible to eliminate the series resistances and to compensate for pressure variations by the control [87].

It was checked experimentally that supplying a laser with a current having a high ripple content at twice the switching frequency did not affect its operation and that the laser behaved as an ideal voltage source at high frequencies. It has even been observed that the quality of the laser beam at switch-on and at low ionizing currents was improved.

The complete power circuit is given in Fig. 7.13. Each gun is fed by a rectifier and a transformer, and all the primaries are connected in series with the resonant network. The ionizing current, measured by a single shunt at the output, is controlled by varying the frequency of the inverter.

The prototype, based on two inverter legs such as those described in Section 7.3.4, could attain the nominal point of operation when supplied by the 380 V/50 Hz mains with a current in the resonant network of less than 30A. On-off operation with a 50% duty cycle and a frequency in excess of 3 kHz demonstrated the flexibility of the control in this power supply. The weight and size of the whole power supply was divided by four which allowed integration of the supply into the laser apparatus.

This power supply is now being engineered for manufacture.

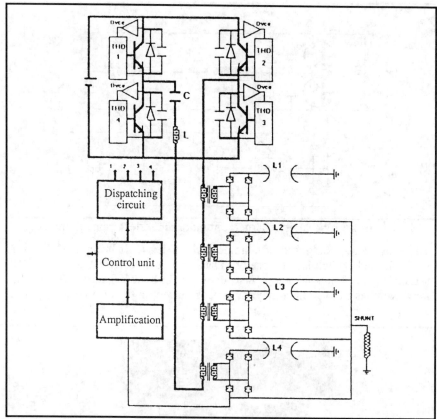

Fig. 7.13 Circuit of the power supply for the four-beam CO_2 laser with control arrangement

7.3.4 Battery charger

The battery charger developed in collaboration with the T.E.G. company (previously A.T.E.I.) is operated from the 220 V/50 Hz mains and can deliver 1.5 kW at 48 V. The structure of this resonant battery charger is shown in Fig. 7.14 ·[94,100].

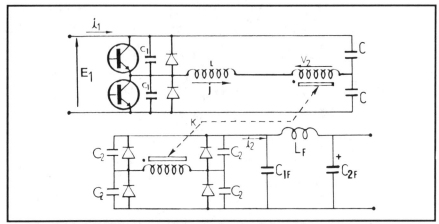

Fig. 7.14 1.5 kW battery charger: circuit of the series resonant inverter

The inverter is composed of a single inverter leg (half-bridge) as shown in Fig. 7.15. Each switch is composed of Darlington-connected BUS 48 and BUS 98 transistors (Thomson) and a fast recovery BYX 62600 diode (Thomson). These switches can handle a maximum peak current in the resonant network of 30 A.

Fig. 7.15 1.5 kW battery charger: practical implementation of the inverter

A snubber capacitor C_1 of 22 nF is connected in parallel with each switch. Hence, when operating at 25 kHz, with an input voltage of 300 V, and a switched current of 25 A, the conduction losses are 25 W while the switching losses are approximately 1 W.

The resonant network is composed of an inductor L of 90 µH and of two capacitors C of 680 nF that also act to provide a capacitive centre tap for the inverter. The step-down transformer has a ratio of 2:1 and the value of the capacitors C_2 is 1.22 µF.

Although the output voltage is relatively low, a full bridge rectifier was chosen. Tests of centre tapped rectifiers, a priori more suitable for low voltage operation, revealed unacceptable oscillations of the secondary current because of the capacitors C_2 and the leakage inductance of the secondary windings.

The structure and design of the output filter results mainly from a compromise between dynamic performance and acceptable output voltage ripple. In order to obtain a sufficient bandwidth, a third order filter was selected. The peak-to-peak output voltage ripple is approximately 100 mV.

A control diagram for this charger, designed to allow easy paralleling of several chargers of this type, is shown in Fig. 7.16. Each charger is controlled to give current source behaviour, and when paralleled, the voltage regulator provides a common current reference which ensures correct sharing of the load current. It should be noted here that this regulation is easily achieved since the resonant converter naturally behaves as a current source.

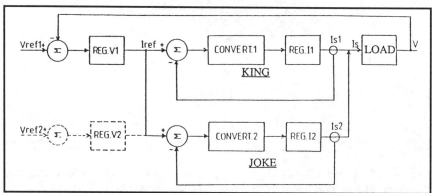

Fig. 7.16 1.5 kW battery charger: control diagram and parallel operation

The charger comes in a 3U/19 inch standard rack (d:360 mm w:430 mm h:115 mm) and the overall weight is 12 kg. The efficiency is greater than 90% at full power. As far as the conducted and radiated noise is concerned, this charger conforms to the following standards: VDE 871 Class B, military standard GAM T 13, and the CNET specifications.

The T.E.G. company has also realized prototypes of 3 kW and 6 kW chargers based on the same principle. Because of its topology the 6 kW is the more attractive with respect to the power/volume ratio and price.

7.3.5 Three phase sinewave inverter

A 1 kVA 115/200 V three phase sinewave inverter with very low harmonic distortion and output frequency variable from 0 to 1 kHz was also developed in collaboration with the same company, T.E.G. [94,101].

The structure of this inverter is indicated in Fig 6.48 (differential conversion). Converters CS_1 and CS_2 are half bridges and the power block (Fig. 7.17) is made up of five dual thyristor inverter legs such as those of Fig 7.15. The switching frequency is set at 40 kHz.

The inductor of the resonant network and the four high frequency transformers are located in a cooling duct made up by the heat sinks of the five inverter legs and the sides of the apparatus (Fig. 7.17). Such an arrangement allows simultaneous cooling of the heatsinks, the inductor and the transformers. Furthermore, this high frequency circuit is thus screened by this cooling duct which reduces the electromagnetic radiation.

Fig. 7.17 Breadboard of the 1 kVA three-phase sinusoidal inverter

This three phase inverter weighs less than 22 kg for a volume less than 33 l, is capable of supplying both inductive and capacitive loads with no power factor limitation, and can supply totally unbalanced loads.

The stability of the output voltage is ±0.5% while distortion of the output voltages is less than 0.5% for frequencies up to 500 Hz and less than 2% for frequencies up to 1 kHz. Efficiency in excess of 75% is achieved.

Conformity to the usual electromagnetic noise standards (V.D.E. 871B and C.I.S.P.R., B curve) has easily been attained by using small mains filters and by obeying the usual layout rules for high frequency switching converters. This apparatus is presently being successfully marketed, as well as a 3 kVA variant.

7.4 CONCLUSION

After outlining different methods that provide more or less accurate approaches to the design of a resonant converter, some convincing examples of the application of resonant conversion have been given, all of which have given rise to industrial products and even their commercial exploitation.

Each of these applications is very specific and demands features of the converter performance that only resonant techniques can provide, most especially together.

Studies concerning other applications have also been carried out:
- electric soldering (S.A.F.),
- cutting out of sheet metal by means of a plasma torch (S.A.F.),

or are currently being carried out:
- AC/DC conversion (Auxilec),
- asynchronous motor drives (Technicatome).

Although resonant converters can fulfill very demanding specifications, their structures are complex and their design requires a thorough knowledge of their operation. This is almost certainly the reason why resonant converters are only used in high technology fields such as aeronautics and military or medical electronics.

8
CONCLUSION

In a static converter, the switches are clearly the elements that are subject to the most severe stresses precisely because of their operation as switches. Increasing the switching frequency requires a reduction in the stresses applied to the switches during the commutations to lower the losses and to increase the efficiency but also to improve the reliability of the equipment.

Thus, one way to improve the performance of the converters is clearly to employ **soft commutation,** that is to say the utilization of switches that all have at least one inherent commutation and at most one controlled commutation.

Their zero current or zero voltage inherent switchings are naturally lossless. Furthermore, the controlled commutation of these switches can be effectively protected by non-dissipative snubbers that significantly reduce the losses.

From the converter viewpoint, the two conditions for soft commutation can be stated as follows:

1. At least one of the voltage and current sources must be reversible — voltage-reversibility if a voltage source and/or current-reversible if a current source.
2. The switching frequency of the switches of the cell must be tightly linked — and generally equal — to the frequency of the changes of sign of the alternating quantity that yields the soft commutation.

When the conditions for soft commutation are naturally met in a converter, the commutation is called **natural commutation.**

In other cases, the converter must be equipped with an auxiliary network composed of inductors and capacitors arranged to enforce the soft commutation conditions whatever the conditions imposed by the sources.

This auxiliary network can ensure soft commutation of the switches locally. This method is not really new since it has been widely used when the only switching devices available were thyristors and diodes. Such a technique is referred to as forced commutation. Though forced commutation is traditionally associated with thyristors, it is also suitable for dual thyristors.

Even if the commutation problems are a priori solved with the fully controlled switches currently available, forced commutation becomes topical again with the quasi-resonant converters. Because of the intrinsic speed of modern components and of the reduction of the switching losses, the commutation circuit benefits from the reduction in size following from the increase in frequency. Under such circumstances these circuits, which also possess quasi sinusoidal waveforms, can be an attractive alternative to the switch-mode converters.

Among the numerous commutation circuits which have been presented, some are not totally new but others are more original and join together switches with commutation mechanisms which are the duals of each other. These circuits merit special attention now that the thyristor (turn-on controlled device with inherent turn off) and the GTO operated in the dual thyristor mode (turn-off controlled device with inherent turn on) can commutate powers in the same range.

The auxiliary reactive network can also be an AC intermediate stage that globally ensures soft commutation of the converters it is connected to. This second technique generally leads to AC link converters of which resonant converters are evidently the major group.

The characteristics and properties of the non-reversible DC/DC resonant converters have been demonstrated and the control laws have been listed. Their principles have then been generalized to all the major types of electrical energy conversion.

The technology of the series resonant converter, using switches which are not voltage-reversible, yields the best frequency response. Furthermore, such converters are entirely compatible with the reality of imperfect sources: a DC source can readily be decoupled and regarded as a perfect voltage source whilst an AC source, which usually has some inductive impedance that readily be supplemented, can be regarded as a current source.

Besides realising soft commutation, resonance is a technique that also enables selection of the most suitable type of commutation and thus make the best use of modern power semiconductors.

Turn-off controlled commutation (dual thyristor) is the most suitable for application with series resonance. In particular, it takes advantage of the parasitic capacitance of the semiconductor devices, which then act as snubber capacitors, it alleviates the problems associated with diode turn-off and results in highly reliable operation as possible short-circuits through the inverter legs are naturally precluded.

PWM converters, like series resonant converters, are in most cases supplied with a DC voltage source and in nearly all applications must behave as voltage sources at least as regards their instantaneous behaviour. This is naturally obtained with the series resonant converter because of the single filter capacitor and requires an LC filter in PWM converters. A direct consequence is that series resonant converters have far better dynamic behaviour than their PWM counterparts.

In the special case of supplying a non-linear AC load, the LC filter of a PWM inverter plays a vital role because it performs two conflicting duties; on the one hand it filters the harmonics generated by the switching, and on the other hand, it filters the current harmonics generated by the load and distorting the terminal voltage [102~104].

Removing the filter, or at least placing its cut-off frequency well beyond the harmonics of the fundamental output frequency solves this problem and greatly simplifies the control. The voltage across the load only has to be controlled with respect to a sinusoidal reference that is varying slowly with respect to the switching period.

In order to achieve this, however, it is necessary to reach switching frequencies much higher than the highest harmonic frequencies. Given the state-of-the-art technology of power semiconductors, this seems to be attainable only with the use of soft commutation techniques and of resonant converters that fundamentally solve the special problem of supplying non-linear loads.

In addition to this property, the results already obtained with a number of industrial products augur well for the large-scale development of these resonant conversion systems and fully justify the varied research programs being carried-out in the laboratories of both universities and companies.

Lastly, because of our normal working environment, it is difficult to conclude without speaking of the teaching implications [105].

Given the variety and complexity of the converters studied, we have been led to formulate precise statements of some of the basic principles of power electronics, especially those related to sources, switches and power transfer reversibility. Disregarding the often complicated second order phenomena, these principles allow a clear elucidation of the converter operation and thus enable a configuration to be established.

The concepts of duality and commutation cells have also been defined.

Duality applied to the structures and/or to the commutation mechanisms of the switches is a powerful and indispensable tool in making a clear and synthetic presentation of power electronic circuits.

In a converter, each switch belongs to at least one commutation cell. The functioning of this cell results in full specifications for the switches it is made of, regarding their static characteristics as well as their dynamic characteristics.

The commutation cell concept is a very important concept since it allows establishing a relation between the structure of a converter (or its function) and the technology. This link between two points of view, which are so often set apart, also leads to the conception of fruitful research projects involving collaboration between component specialists and circuit designers.

As we hope to have demonstrated in this book, these tools are a considerable aid to creativity. Mastering the various essential elements also enables a logical and systematic approach to the synthesis and presentation of static converters. power electronics thus acquires an outline theoretical basis, and the teaching of it need no longer consist of a series of unrelated case studies.

APPENDIX A1
NOTATION -
NORMALIZED UNITS

The different analytical studies of the series resonant converter (Fig. A1.1) presented in this book use a system of base units consisting of:
• input voltage E_1 when the inverter is a full-bridge, or half the input voltage when the inverter is a half-bridge,
• $\sqrt{L/C}$, the characteristic impedance of the resonant network,
• current $I_B = E_1\sqrt{C/L}$,
• power $E_1^2\sqrt{C/L}$,
• the natural period of the resonant network: $T_0 = 2\pi\sqrt{LC}$,
• the natural frequency of the resonant network: $f_0 = 1/T_0$.

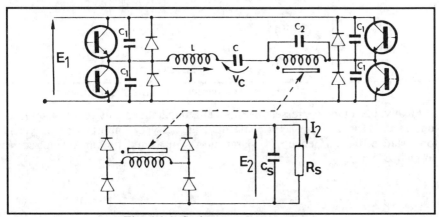

Fig. A1.1 Series resonant converter

From this system, a set of normalized units can be defined:
• $y = j/I_B$: normalized current in the resonant network,
• $x = v_c/E_1$: normalized voltage across capacitor C,
• $Y_{avg} = I_2/I_B$: normalized average current in the load,
• $q = E_2/E_1$: normalized voltage across the load,
• $u = f_s/f_0$ where f_s is the frequency of operation of the converter.

In an analytical study, these normalized quantities appear naturally and lead to the derivation of dimensionless equations that characterize a structure instead of a particular circuit.

Furthermore, these normalized quantities are well suited to digital simulation provided a normalized circuit like that of Fig. A1.2 is simulated. The resonant network is composed of a 1 H inductor and a 1 F capacitor. The capacitative snubbers are respectively $\alpha_1 = C_1/C$ and $\alpha_2 = C_2/C$, the frequency of operation is $u = f_s/f_0$ and the input and output voltages are respectively 1 V and q V assuming a 1:1 transformer. Under these conditions, the synthesis element of the analytical study is preserved within the digital simulation.

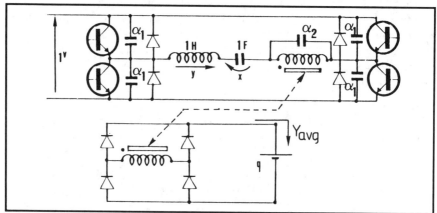

Fig. A1.2 Series resonant converter: normalized quantities

Finally, it should be stressed that normalized quantities only remain useful so long as the base units are kept constant. This is a very important point when, for instance, the influence of a parameter is measured by curves plotted in normalized units.

APPENDIX A2

SMALL SIGNAL ANALYSIS AND MODELLING OF THE SERIES RESONANT CONVERTER OPERATING ABOVE THE NATURAL FREQUENCY

A2.1 ASSUMPTIONS

The diagram of the converter under consideration is shown in Fig. A1.1. The following assumptions are made:
• The switches are ideal: the switching times, the ON-state voltage drop and the leakage current in the OFF-state are neglected,
• the input and output voltages are perfectly smooth,
• the resonant network is lossless,
• the transformer is ideal and has a 1:1 ratio, and
• the effect of the snubbers and of the protection circuits of both the inverter and the rectifier are neglected.

A2.2 STATE PLANE ANALYSIS

The operation of the converter consists of four modes and the corresponding circuits are indicated in Fig. A2.1.

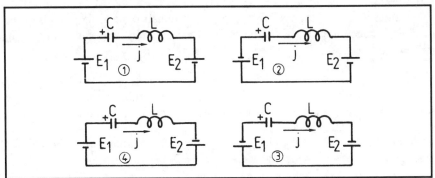

Fig. A2.1 Sequence of modes of operation
of an ideal series resonant converter ($f_s > f_0$)

169

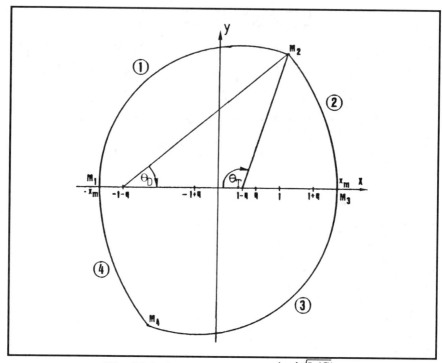

Fig. A2.2 State plane analysis $(v_c, i_1\sqrt{L/C})$
of an ideal series resonant converter

Using the normalized quantities defined in Appendix A1, the state plane analysis of the operation of the series resonant converter operating above the natural frequency is indicated in Fig. A2.2.

θ_T and θ_D are the conduction angles of the transistors and the diodes of the inverter, respectively. x_m is the normalized peak voltage across capacitor C of the resonant network:

$$x_m = V_{Cmax}/E_1 \qquad (A2.1)$$

Points M_2 and M_4, symmetrical about the origin, correspond to the commutations of the inverter. In steady state operation these points are always in the first and third quadrants.

Points M_1 and M_3 at x_m and $-x_m$ represent the commutations of the diode rectifier. These points are always located on the X-axis.

A2.3 STUDY OF THE STEADY STATE MODE

From the state plane analysis, we infer directly:

$$x_m = \frac{(1-q)(1-\cos \theta_T)}{q+\cos \theta_T}$$

(A2.2)

$$x_2 = qx_m = \frac{q(1-q)(1-\cos \theta_T)}{q+\cos \theta_T}$$

(A2.3)

$$y_2 = \frac{(1-q^2) \sin \theta_T}{q+\cos \theta_T}$$

(A2.4)

Also:

$$u = \frac{\pi}{\theta_T+\theta_D}$$

(A2.5)

and

$$Y_{avg} = \frac{2u}{\pi}x_m$$

(A2.6)

In short-circuit operation ($q = 0$), the commutation points M_2 and M_4 move on to the Y-axis and we have:

$$\theta_T = \theta_D = \delta$$

(A2.7)

where:
$$\delta = \pi /2u$$

(A2.8)

The switches then commutate the peak current of the resonant network.

When operating at the natural frequency ($q = 1$), commutation points M_2 and M_4 move along the X-axis. Only the transistors of the inverter are conducting and commutate at the current zero-crossing. This operation corresponds to a theoretical limit for the commutation of the dual thyristors.

The relations established previously lead directly to a plot of the output characteristics $q(Y_{avg})$ corresponding to fixed conduction angles of transistors θ_T (Fig. A2.3).

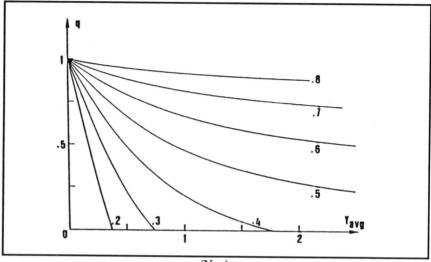

Fig. A2.3 Output characteristics $q(Y_{avg})$ with fixed conduction time for the transistors and for different values of parameter θ_T/π

A2.4 STATIC CHARACTERISTICS WITH FIXED u

From the state plane analysis of Fig. A2.2:

$$\tan \theta_D = \frac{y_2}{x_2+1+q}$$

$$(A2.9)$$

Substituting for x_2 and y_2, we get:

$$q = \frac{\sin \theta_T - \sin \theta_D}{\sin 2\delta}$$

$$(A2.10)$$

Substituting this expression for q in the equation of x_m, we get:

$$x_m = -1 + \cos \theta_T + \tan \delta.\sin \theta_T$$

$$(A2.11)$$

and thus:

$$\delta Y_{avg} + 1 = \cos \theta_T + \tan \delta.\sin \theta_T$$

$$(A2.12)$$

The expression for q can be rewritten as:

$$q = -\cos\theta_T + \frac{\sin \theta_T}{\tan \delta}$$

$$(A2.13)$$

that is to say:

$$q \tan \delta = \sin \theta_T - \tan \delta \cos\theta_T$$

$$(A2.14)$$

172

Using equations (A2.12) and (A2.14), an analytic expression for the function $q(Y_{avg})$ with fixed u can be established:

$$(q \tan \delta)^2 + (\delta Y_{avg} + 1)^2 = \frac{1}{\cos^2 \delta} \qquad (A2.15)$$

or:

$$(q \sin \delta)^2 + ((\delta Y_{avg} + 1) \cos \delta)^2 = 1 \qquad (A2.16)$$

which is the equation of an ellipse. These characteristics are plotted in Fig. A2.4.

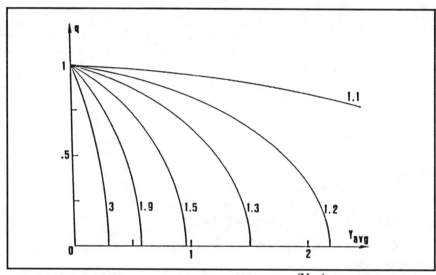

Fig. A2.4 Constant frequency output characteristics $q(Y_{avg})$ for different values of parameter f_s/f_0

Let

$$\rho = R_s \sqrt{C/L} \qquad (A2.17)$$

and

$$K = x_m/q = \delta/\rho \qquad (A2.18)$$

It then becomes possible to establish an expression for the voltage source as a function of the control parameter u and of the load resistance R_s:

$$(q \sin \delta)^2 + ((K q + 1) \cos \delta)^2 = 1 \qquad (A2.19)$$

or:

$$q = \frac{-K.\cos^2 \delta + \sqrt{K^2 \cos^2 \delta + \sin^4 \delta}}{K^2 \cos^2 \delta + \sin^2 \delta} \qquad (A2.20)$$

These q(u) characteristics with constant R_S are given in Fig. A2.5.

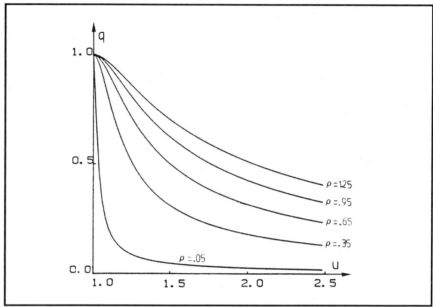

Fig. A2.5 Output voltage versus control parameter u (fixed load)

A2.5 SMALL SIGNAL MODELLING

Using the same assumptions, we wish to define a transfer function for the resonant converter.

Because of the non-linearities of the system, we look for a modelling technique suitable for small signal perturbations, which means low amplitude slowly varying perturbations applied to the converter during normal operation. At any time, the resonant network is in the steady state mode and during the application of the perturbation, the steady state is slowly varying.

The method finally used involves linearizing the characteristics about the point of operation.

Differentiation of Equation (A2.16) gives a relation of the form:

$$A \, dq + B \, dY_{avg} + D \, du = 0 \qquad (A2.21)$$

where:

$$A = 2 q \sin^2\delta \tag{A2.22}$$

$$B = 2\delta\left(1 + \delta\ Y_{avg}\right)\cos^2\delta \tag{A2.23}$$

$$D = \frac{\delta^2 Y_{avg}^2\left(\delta\sin\ 2\delta - 2\cos^2\delta\right) + 2\delta Y_{avg}\left(\delta\sin\ 2\delta - \cos^2\delta\right) + \delta\sin\ 2\delta\left(1 - q^2\right)}{u} \tag{A2.24}$$

From the circuit of Fig. A1.1 and assuming the input voltage to be constant, we get:

$$dq = dY_{avg}\ \frac{\rho}{1 + \tau p} \tag{A2.25}$$

where $\qquad\qquad\qquad\qquad \tau = R_S C_S \tag{A2.26}$

The expression for the small signal transfer function that relates variations of the output voltage dq to variations of the frequency du can be written as follows:

$$\frac{dq}{du} = \frac{G}{1 + \dfrac{p}{\omega_c}} \tag{A2.27}$$

where

$$\omega_c = \frac{1}{\tau}\left(1 + \frac{q\tan^2\delta}{K\left(1 + qK\right)}\right) \tag{A2.28}$$

and

$$G = \frac{1}{u}\ \frac{q + \delta\dfrac{q^2\left(1 + qK\right)^2}{K\left(1 + qK\right)}\tan\ \delta}{1 + \dfrac{q}{K\left(1 + qK\right)}\tan^2\ \delta} \tag{A2.29}$$

The gain and cut-off frequency of this normalized transfer function with respect to u, with ρ as a parameter, are given in Figs A2.6 and A2.7.

The real transfer function reads:

$$\frac{dE_2}{df_s} = \frac{E_1}{f_0}\ \frac{dq}{du} \tag{A2.30}$$

175

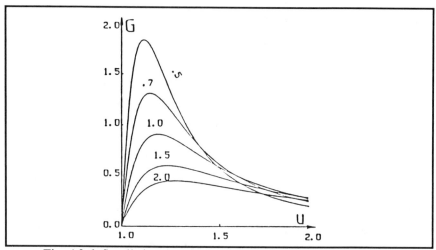

Fig. A2.6 Small signal analysis of the series resonant converter:
gain versus control parameter (fixed load)

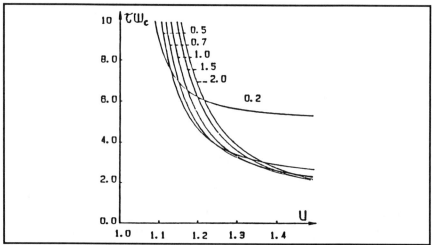

Fig. A2.7 Small signal analysis of the series resonant converter:
cut-off frequency versus control parameter (fixed load)

APPENDIX A3

DETAILED ANALYSIS OF THE SERIES RESONANT CONVERTER OPERATING ABOVE THE NATURAL FREQUENCY

A3.1 INFLUENCE OF THE CAPACITATIVE SNUBBERS IN THE INVERTER

The theoretical analysis of the operation of the resonant converter carried out in Appendix A2 has been experimentally checked on a converter of a few kVA. The transformer was removed to allow the assumption of a perfect transformer with a 1:1 ratio. The different experiments conducted with this converter showed that the theoretical analysis gave a good description of the operation though it did not show one important phenomenon: at constant frequency when a characteristic is followed starting from the short-circuit mode, increasing the load resistance always eventually leads to a commutation failure in the inverter. This commutation failure occurs sooner as the capacitative snubbers get larger. This observation lead to a new analysis of the converter involving the snubbers in the inverter.

The study of the resonant converter presented here is carried out under the following assumptions:
• The transformer is ideal and has a 1:1 turns ratio.
• The commutations of the semiconductors are instantaneous.
• The DC voltage ripple across the inverter and the load is negligible.
• The losses in the resonant network are negligible.

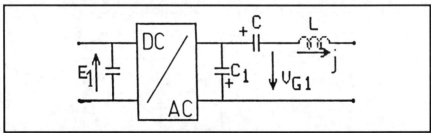

Fig. A3.1 Equivalent circuit for analysis including the snubber capacitors in the inverter

Taking the snubbers of the inverter (Fig. A1.1) into account amounts to consider a single capacitor C_1 parallel connected to the output of the converter (Fig. A3.1). Let us recall here that this is feasible because the switches turn on inherently at zero voltage instead of being turned on at any time (dual thyristor).

Under such circumstances, the operation of the converter can be split into six modes (Fig. A3.2). Though the commutations of the switches are instantaneous, the commutations of the inverter output voltage are not instantaneous anymore. During these commutations, no switching device in the inverter is on and current j in the resonant network flows through capacitor C_1 (modes 2,5).

Let: $$C_1 = a_1 C \qquad (A3.1)$$

and $$k_1{}^2 = (a_1 + 1)/a_1 \qquad (A3.2)$$

Each circuit of Fig. A3.2 is composed of a resonant network with infinite Q fed by a voltage source. The state plane analysis is thus possible, though only the quantities related to a single resonant network can be drawn on the same state plane. Thus, in order to carry out this study, two state planes should be used, one to show the operation between commutations (modes 1,2,3,4) and the other to represent the commutations (modes 2 and 5).

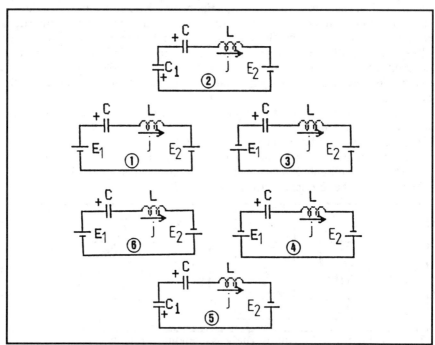

Fig. A3.2 Analysis including the snubber capacitors of the inverter:
sequence of modes of operation

Nevertheless, the study of these commutations when current j is positive (Fig. A3.2, mode 2), shows that it is possible to use a single state plane to represent all the operation of the converter.

Let G_1 be the capacitor equivalent to C in series with C_1:

$$G_1 = C_1 C/(C + C_1) \qquad (A3.3)$$

Using suffix B to denote the values of the different quantities at the beginning of the commutation and suffix E to denote the values at the end of the commutation, we get:

$$v_{C1E} = v_{C1B} + 2E_1 \qquad (A3.4)$$

$$v_{CE} = v_{CB} + 2a_1 E_1 \qquad (A3.5)$$

$$(v_{G1B} + E_2)^2 + (j_B \sqrt{L/G_1})^2 = (v_{G1E} + E_2)^2 + (j_E \sqrt{L/G_1})^2 \qquad (A3.6)$$

Using Equations (A2.4),(A2.5) and (A2.6), we get:

$$(v_{CB} + E_2)^2 + (j_B \sqrt{L/C})^2 = (v_{CE} + E_2)^2 + (j_E \sqrt{L/C})^2 \qquad (A3.7)$$

which can be rewritten as a function of the normalized quantities:

$$(x_B + q)^2 + y_B{}^2 = (x_E + q)^2 + y_E{}^2 \qquad (A3.8)$$

This implies that the points that represent the end of mode 1 and the beginning of mode 3 in the (x,y) state plane both belong to an arc of a circle centred at (-q,0). It should be noted that on this circular arc only the extreme points are significant. Furthermore, the angle at the centre of this arc is also not significant, since the electrical quantities vary with angular frequency ω_1 (natural angular frequency of the LG_1 network) rather than angular frequency ω_0.

With these comments taken into account, the state plane analysis of the operation of the converter, including the influence of the capacitative snubbers in the inverter is given in Fig. A3.3.

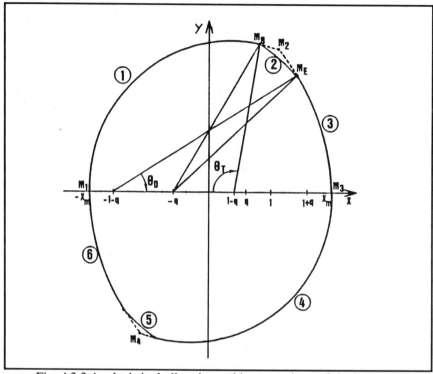

Fig. A3.3 Analysis including the snubber capacitors of the inverter:
representation of the operation in the state plane

The other relations that enable full determination of the state plane trajectories are:

$$(x_B - 1 + q)^2 + y_B{}^2 = (-x_m - 1 + q)^2 \qquad (A3.9)$$

$$(x_E + 1 + q)^2 + y_E{}^2 = (x_m + 1 + q)^2 \qquad (A3.10)$$

From (A3.8): $\qquad y_E{}^2 = y_B{}^2 - 4a_1(x_B + a_1 + q) \qquad (A3.11)$

Replacing x_E and y_E with their values given by (A3.10), and subtracting (A3.10) from (A3.9) we get:

$$x_B = qx_m - a_1 \qquad (A3.12)$$

$$x_E = qx_m + a_1 \qquad (A3.13)$$

These two results agree with the fact that, if the capacitative snubbers are reduced — that is to say if a_1 approaches 0 — then x_B and x_E should tend to qx_m (c.f. Appendix A2).

180

Using the state plane analysis (Fig. A3.3) and as functions of θ_T, which is the normalized conduction time of the transistors in the inverter, we can also derive the following:

$$x_m = \frac{a_1 + (1-q)(1-\cos\theta_T)}{q + \cos\theta_T} \tag{A3.14}$$

$$y_B = (x_m + 1 - q)\sin\theta_T \tag{A3.15}$$

$$\theta_c = \frac{1}{k_1}\left(\tan^{-1}\frac{k_1 y_B}{x_B + q - 1} - \tan^{-1}\frac{k_1 y_E}{x_E + q - 1}\right) \tag{A3.16}$$

$$\theta_D = \tan^{-1}\frac{y_E}{x_E + q - 1} \tag{A3.17}$$

In these equations θ_c and θ_D are respectively the normalized commutation times and the conduction times of the inverter diodes and function $\tan^{-1}(x)$ is the inverse tangent of x with values lying between 0 and π.

The average load current is:

$$Y_{avg} = 2ux_m/\pi \tag{A3.18}$$

where:
$$u = \frac{\pi}{\theta_T + \theta_c + \theta_D} \tag{A3.19}$$

The commutation of the dual thyristor inverter is guaranteed if the bias of capacitor C_1 is reversed before current j in the resonant network vanishes. Then we get:

$$x_m > a_1/(1 - q) \tag{A3.20}$$

which amounts to:

$$Y_{avg} > 2ua_1/\pi(1 - q) \tag{A3.21}$$

The constant frequency output characteristics $q(Y_{avg})$ showing this commutation limit are shown in Fig. A3.4.

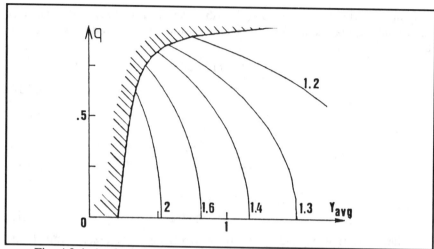

Fig. A3.4 Analysis including the snubber capacitors of the inverter:
output characteristics

A3.2 INFLUENCE OF TRANSFORMER IMPERFECTIONS

The construction of a breadboard circuit using a transformer with a high turns
ratio showed that the analysis had to be still more detailed to give an accurate
explanation of the experimental results. Simply adding the step-up transformer
enables certain regions of the no-load operation, to such an extent that the
commutation limit created by the snubbers in the inverter seems to disappear
completely. Amongst all the assumptions, the one stating that such a step-up
transformer is perfect is obviously an oversimplification. Thus an equivalent
circuit of this transformer has been derived from an analysis of the primary
voltage waveform (Fig. A3.5).

Commutation of the inverter generates a voltage step across the primary
because of the leakage inductance. The amplitude of this step depends on the
voltage divider composed of the inductance in the resonant network and of the
total leakage inductance of the primary N_1:

$$\Delta V_P = 2E_1 N_1/(L + N_1) \qquad\qquad (A3.22)$$

At the zero crossing of current j in the resonant network, the commutation of
the voltage on the secondary side of the transformer should be instantaneous and
thus generate a significant dV/dt. The experimental waveform (Fig. A3.5) does
not show these properties, which can be explained by the presence of parasitic
capacitance connected in parallel to the secondary winding of the transformer.

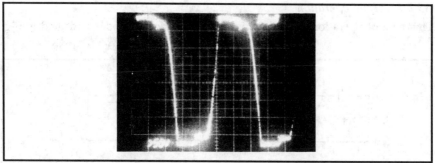

Fig. A3.5 Experimental waveform of the primary voltage across a high-ratio
step-up transformer in a series resonant converter

Since the magnetizing current is negligible compared with the current in the
resonant network, the circuit of Fig. A3.6 is a suitable equivalent circuit of the
transformer. It is composed of the total primary leakage inductance N_1, the
parasitic capacitance referred to the primary side C_2 and an ideal transformer. The
leakage inductance N_1 is in series with the inductance of the resonant network
and can be assumed to be a fraction of the latter. Thus, this leakage inductance is
not responsible for the change in the operating mode of the converter and will
not be considered in the following study. The really new element of this
equivalent circuit is capacitor C_2.

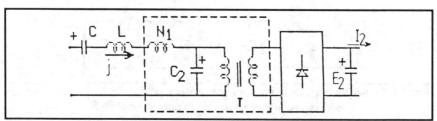

Fig. A3.6 Model for the step-up transformer

Let:
$$C_2 = a_2 C \qquad (A3.23)$$

$$k_2^2 = (a_2 + 1)/a_2 \qquad (A3.24)$$

$$G_2 = C_2 C/(C + C_2) \qquad (A3.25)$$

and to keep the expressions simple, the ideal transformer of Fig. A3.6 is
assumed to have a 1:1 turns ratio.

It should be noted that capacitor C_2 also acts as a snubber capacitor for the
rectifier diodes. For this reason this capacitor may exist even without a
transformer with a high step-up ratio. Also, under no-load condition ($I_2=0$), it
enables a certain current j to flow which allows operation of the dual thyristor
inverter in the frequency range above the natural frequency of the LG_2 network.

A3.2.1 Normal operation

This type of operation is characterized by the series of modes indicated in Fig. A3.7.

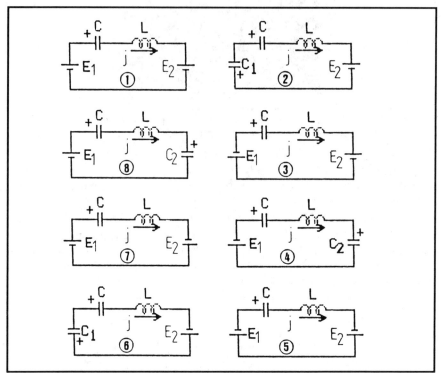

Fig. A3.7 Analysis including the snubber capacitors of both the inverter and the rectifier: sequence of modes in normal operation

Modes 2 and 6, related to the commutation of the output voltage of the inverter, are initiated by control signals. By contrast, modes 4 and 8, which correspond to the commutations of the rectifier are initiated by the zero crossing of current j in the LC resonant network. All these modes come to an end when the bias on one or other of the capacitors C_1 and C_2 is reversed.

Phases 1,4,5 and 8 represent the conduction of the transistors of the inverter while phases 3 and 7 represent the conduction of the inverter diodes.

In the light of the remark made in the preceding paragraph, the operation can be described in the state plane with mode i being represented by the arc of a circle $M_i M_{i+1}$ (Fig. A3.8). Only modes 1 to 4 need be analysed because the operation is symmetrical, and thus in the steady state conditions:

$$x_5 = -x_1 \tag{A3.26}$$

$$y_5 = -y_1 \tag{A3.27}$$

The key relationships that fully determine the state plane trajectories are:

$$(x_1-1+q)^2 + y_1{}^2 = (x_2-1+q)^2 + y_2{}^2 \qquad (A3.28)$$

$$(x_2+q)^2 + y_2{}^2 = (x_3+q)^2 + y_3{}^2 \qquad (A3.29)$$

$$(x_3+1+q)^2 + y_3{}^2 = (x_4+1+q)^2 + y_4{}^2 \qquad (A3.30)$$

$$(x_4+1)^2 + y_4{}^2 = (x_5+1)^2 + y_5{}^2 \qquad (A3.31)$$

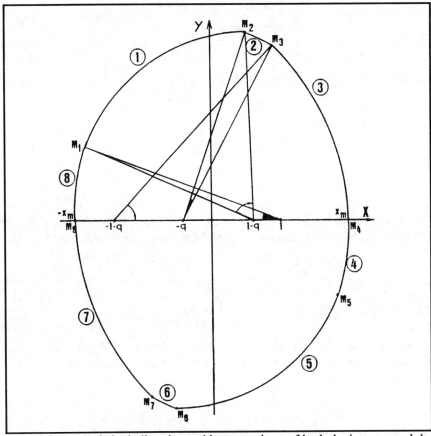

Fig. A3.8 Analysis including the snubber capacitors of both the inverter and the rectifier: state plane trajectory in normal operation

During the commutations of the inverter and of the rectifier, the voltage variations across capacitor C are given by $2a_1E_1$ and $2a_2E_2$ respectively and consequently:

$$x_3 = x_2 + 2a_1 \tag{A3.32}$$

$$x_5 = x_4 - 2a_2q \tag{A3.33}$$

Equations (A3.26) to (A3.33) lead to all the coordinates being expressed as functions of x_m, q, a_1 and a_2. Only the abscissa of points M_i are given here since they are those used particularly in the following study:

$$x_1 = -x_m + 2a_2q \tag{A3.34}$$

$$x_2 = qx_m - a_2q^2 - a_1 \tag{A3.35}$$

$$x_3 = qx_m - a_2q^2 + a_1 \tag{A3.36}$$

$$x_4 = x_m \tag{A3.37}$$

The normalized average load current is given by:

$$Y_{avg} = 2u\ (x_m - a_2q)/\pi \tag{A3.38}$$

and the ratio $u = f_s/f_0$ by:

$$u = \frac{\pi}{\displaystyle\sum_{i=1}^{4} \omega_0 t_i} \tag{A3.39}$$

where t_i is the duration of mode i.

For the modes 1 and 3 that involve circuits with angular frequency ω_0, their durations can be directly derived from the state plane (Fig. A3.8):

$$t_1 = \frac{1}{\omega_0}\left(\tan^{-1}\frac{y_1}{x_1 - 1 + q} - \tan^{-1}\frac{y_2}{x_2 - 1 + q}\right) \tag{A3.40}$$

$$t_3 = \frac{1}{\omega_0}\tan^{-1}\frac{y_3}{x_3 + 1 + q} \tag{A3.41}$$

Modes 2 and 4 respectively involve circuits with natural angular frequencies ω_1 and ω_2 and thus it is necessary to be careful when determining their durations:

$$t_2 = \frac{1}{\omega_1}\left(\tan^{-1}\frac{k_1 y_2}{x_2 - 1 + q} - \tan^{-1}\frac{k_1 y_3}{x_3 + 1 + q}\right) \tag{A3.42}$$

$$t_4 = \frac{1}{\omega_2}\left(\pi - \tan^{-1}\frac{k_2 y_1}{x_1 - 1 + q}\right) \tag{A3.43}$$

Regions of validity

The mathematical relations (A3.26) to (A3.43) provide an anlysis of the normal operation in steady state of the converter. However the region of validity for these equations corresponding to the sequence of modes exposed in Fig. A3.7 must be defined, using the constraints in (A3.44) and (A3.45):

$$x_1 < x_2 \tag{A3.44}$$

$$x_3 < x_m \tag{A3.45}$$

The constraint (A3.44) means that the commutation of the rectifier finishes before the commutation of the inverter is initiated.

The constraint (A3.45) means that the commutation of the inverter has finished when the current in the resonant network vanishes. This relation is the condition for normal operation of a dual thyristor inverter.

All the variables within the converter can be expressed as functions of x_m, q, a_1 and a_2. Since the coefficients a_1 and a_2 are constant for a given converter, the region of validity in the $q(x_m)$ plane is defined by the following two boundaries:

$$x_1 = x_2 \tag{A3.46}$$

$$x_3 = x_m \tag{A3.47}$$

which lead to:

$$x_m = (a_2 q^2 + a_1 q + a_1)/(q + 1) \tag{A3.48}$$

$$x_m = (a_1 - a_2 q^2)/(1 - q) \tag{A3.49}$$

These boundaries have different shapes depending on the relative values of a_1 and a_2 and are shown in Fig. A3.11.

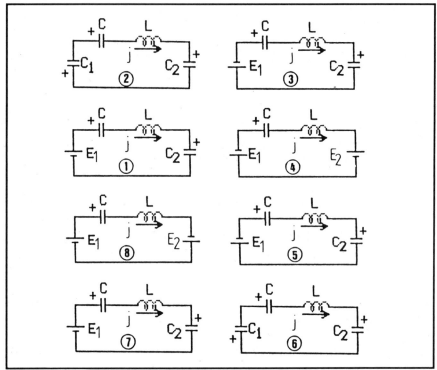

Fig. A3.9 Analysis including the snubber capacitors of both the inverter and the rectifier: sequence of modes in the secondary region of operation

A3.2.2 Secondary region of operation

The secondary region of operation is characterized by the fact that the commutation of the rectifier finishes after that of the inverter. It is represented by the sequence of modes of Fig. A3.9 which involve only undamped second order networks with three different natural frequencies. The state plane representation (Fig. A3.10) thus becomes complicated and requires two different state planes which are, on the one hand, the (x,y) plane we have already used for modes 4 and 8 and, on the other hand, the (x°,y°) plane defined as:

$$x^\circ = v_{G2}/E_1 \qquad\qquad (A3.50)$$

and
$$y^\circ = k_2 y \qquad\qquad (A3.51)$$

for all the other modes in Fig. A3.9.

We will denote by G_{12} the capacitance equivalent to C, C_1 and C_2 connected in series.

$$G_{12} = CC_1C_2/(CC_1 + CC_2 + C_1C_2) \qquad\qquad (A3.52)$$

and we define:

$$k_{12}{}^2 = 1 + 1/a_1 + 1/a_2 \qquad\qquad (A3.53)$$

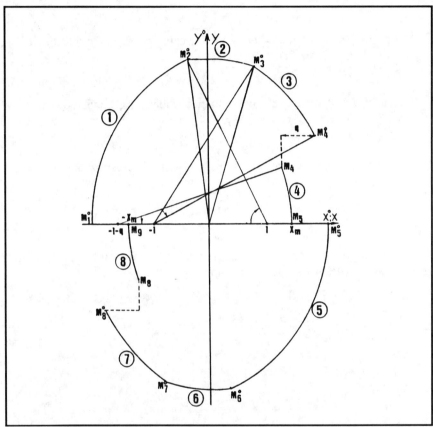

Fig. A3.10 Analysis including the snubber capacitors of both the inverter and the rectifier: state plane trajectory in the secondary region of operation

Using the same conventions as for the normal operation, phase i is represented by the arc of a circle $M°_iM°_{i+1}$ or M_iM_{i+1}. For the reasons given in the preceding paragraph, only the extreme points of the circular arcs $M°_2M°_3$ and $M°_6M°_7$ are significant. The key equation that fully determine the state plane trajectories are the following:

$$(x_m+1+q)^2 = (x°_2+1)^2 + y°_2{}^2 \qquad\qquad (A3.54)$$

$$x°_2{}^2 + y°_2{}^2 = x°_3{}^2 + y°_3{}^2 \qquad\qquad (A3.55)$$

$$(x°_3+1)^2 + y°_3{}^2 = (x°_4+1)^2 + y°_4{}^2 \qquad\qquad (A3.56)$$

$$(x_4+1+q)^2 + y_4{}^2 = (x_m+1+q)^2 \qquad\qquad (A3.57)$$

During the commutation of the inverter, capacitor G_2 undergoes a voltage change of $2E_1C_1/G_2$ so that:

$$x°_3 = x°_2 + 2\, a_1 k_2{}^2 \qquad\qquad (A3.58)$$

Similarly, during the rectifier commutation, capacitor G_2 undergoes a voltage change of $2E_2(1+a_2)$ and consequently:

$$x°_4 = -x_m + q\,(1 + 2a_2) \qquad\qquad (A3.59)$$

Since the current in the inductor L and the voltage across capacitor C cannot change instantaneously, we have the following two relationships:

$$x_4 = x°_4 - q \qquad\qquad (A3.60)$$

and
$$y_4 = y°_4/k_2 \qquad\qquad (A3.61)$$

The other abscissa for the different points M_i or $M°_i$ can be easily inferred from equations (A3.50) to (A3.61) and are:

$$x°_1 = -x_m - q \qquad\qquad (A3.62)$$

$$x_2° = \frac{(x_m - a_2 q)\,(1 + q + a_2 q)}{a_2} - a_1 k_2^2 \qquad\qquad (A3.63)$$

$$x_3° = \frac{(x_m - a_2 q)\,(1 + q + a_2 q)}{a_2} + a_1 k_2^2 \qquad\qquad (A3.64)$$

$$x_5 = x_m \qquad\qquad (A3.65)$$

Though the rectifier delivers energy only during modes 4 and 8, the average current in the load is still defined by relation (A3.38).

The duration t_i of each mode i (i=1,..,4) is defined as follows:

$$t_1 = \frac{1}{\omega_2}\left(\pi - \tan^{-1}\frac{y_2^o}{x_2^o-1}\right)$$

(A3.66)

$$t_2 = \frac{1}{\omega_{12}}\left(\tan^{-1}\frac{k_{12}y_2^o}{k_2(x_2^o-1)} - \tan^{-1}\frac{k_{12}y_3^o}{k_2(x_3^o+1)}\right)$$

(A3.67)

where ω_{12} is the natural angular frequency of network LG_{12}.

$$t_3 = \frac{1}{\omega_2}\left(\tan^{-1}\frac{y_3^o}{x_3^o+1} - \tan^{-1}\frac{y_4^o}{x_4^o+1}\right)$$

(A3.68)

$$t_4 = \frac{1}{\omega_0}\tan^{-1}\frac{y_4}{x_4+1+q}$$

(A3.69)

and the ratio $u = f_s/f_0$ is still determined by relation (39).

(a) Region of validity
As in normal operation, the region of validity for the mathematical equations describing the secondary region of operation of the converter must be defined. These equations are only valid when the inequality constraints (A3.70) and (A3.71) are satisfied:

$$x^o_4 > x^o_3$$

(A3.70)

$$x_4 < x_m$$

(A3.71)

These equations respectively mean that the inverter finishes commutating before the rectifier and that the rectifier delivers a certain current. Equality of x_4 and x_m corresponds to no-load operation of the converter. The boundaries corresponding to equations (A3.70) and (A3.71) and defined respectively by:

$$x_m = (a_2q^2 + 2a_2q - a_1)/(1 + q)$$

(A3.72)

and

$$x_m = a_2q$$

(A3.73)

are shown in Fig. A3.11. The particular operating point corresponding to critical commutation of the inverter under no load conditions is defined by:

191

$$q = a_1/a_2 \qquad\qquad (A3.74)$$

$$x_m = a_1 \qquad\qquad (A3.75)$$

It should be noted that the coordinates of this point satisfy equation (A3.49) which defines the locus of critical commutation of the inverter in normal operation.

A3.2.3 Other regions of operation

There is a third region of operation in the which the rectifier finishes commutating during the commutation of the inverter. This region of operation is nearly impossible to resolve analytically except for the case of critical commutation of the inverter which is still defined by equation (A3.49). In the $q(x_m)$ plane, the area corresponding to this region of operation is thus fully determined by the previous anlysis (Fig. A3.11). Considering that this area is quite narrow in comparison to the others, linear interpolation between the boundaries of this domain gives a full plot of the output characteristics $q(Y_{avg})$.

A3.2.4 Characteristics

The areas of the $q(x_m)$ plane related to each region of operation of the converter are given in Fig. A3.11 for the different cases possible.

(a) $a_1 \geq = a_2$:

Areas (2) and (3) are of little interest since they require pre-established load conditions. Area (1) is limited by the locus of critical commutation of the inverter and by the x_m axis which corresponds to short circuit conditions. In this area the no-load operation is not possible and the output characteristics $q(Y_{avg})$ are similar to those of Fig. A3.4.

(b) $a_1 \leq a_2$:

The whole operating range of the converter is represented by a single region limited by the locus of critical commutation of the inverter, by the line defined by equation (A3.73) which represents no-load operation and by the x_m axis (short circuit operation).

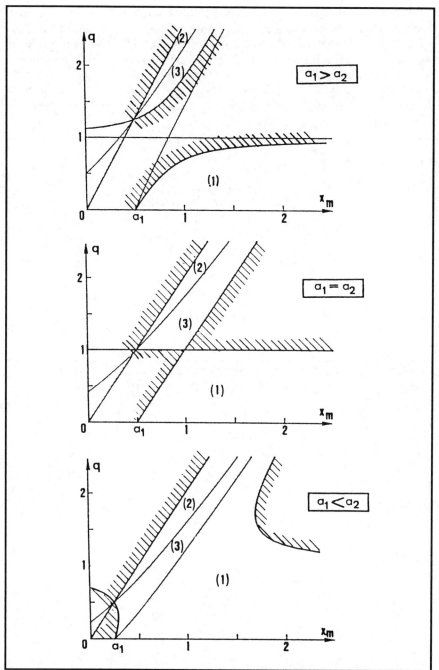

Fig. A3.11 Boundaries of the different modes in the plane $q(x_m)$

Within this area, the constant frequency output characteristics $q(Y_{avg})$ have the appearance of those indicated in Fig. A3.12. Comparing this set of curves to that of Fig. A3.4 reveals the influence of capacitor C_2 on the converter behaviour. Except in the vicinity of the origin, no-load operation is feasible and the commutation problems of the inverter are solved in the area where $q<1$. Generally speaking, C_2 makes the characteristics more vertical, to such an extent that for frequencies in the vicinity of k_2f_0 the series resonant converter behaves almost as a current source.

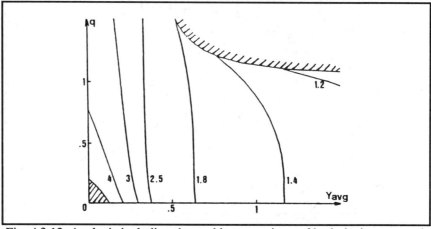

Fig. A3.12 Analysis including the snubber capacitors of both the inverter and the rectifier: output characteristics ($a_1<a_2$)

APPENDIX A4
OPTIMAL CONTROL OF THE SERIES RESONANT CONVERTER

A4.1 INTRODUCTION

In pulse-width-modulated switch-mode converters, the input and output filters usually exhibit time constants far greater than the switching period of the converter. The output quantities are regulated by the direct control of the voltage or current applied to the output filter. Consequently, control techniques are relatively easily put into practice and the dynamic performance, which depends mainly on the characteristics of the filters, can be readily prescribed.

By contrast, a resonant converter involves two types of energy storage circuits, the resonant circuit and the input and output filters which have very different natural modes. The resonant network generally has its natural frequency in the vicinity of the switching frequency while the input and output filters have time constants far greater than the switching period. Furthermore, because of the resonant network, the voltage or the current applied to the output filter is not directly fixed by the control. The energy transfer between the input and the output is achieved through the resonant network which also stores temporarily some energy.

Thus the control of the resonant converter is more complex than that of a switch-mode converter. Transient conditions in a resonant converter such as start-up, shorting of the output or more simply the application of a control step, are rarely predictable and can lead to erratic operation of the converter: commutation failure in the inverter switches, overvoltages and/or overcurrents in the resonant network, etc. Nevertheless, the transient behaviour is strictly dependant on the control strategy chosen for the converter (Section 6.4).

In Appendix A2, a method for studying the steady-state operation of the series resonant converter operating above the natural frequency and based on the state plane analysis (v_C, $j\sqrt{L/C}$) of the resonant network (Fig. A4.1) was presented. The response to slowly varying small signal perturbations — i.e. perturbations such that the resonant network is always in the steady-state mode — were also derived from this state plane analysis. This study enables characterization of the capability of the converter to follow and compensate for slow variations of the control, of the input voltage, or of the load. It also allows provides a basis for stability analysis and the design of the compensating network in the control loop.

195

Still based on the state plane presentation (v_C, $j\sqrt{L/C}$), we will describe a control strategy referred to as optimal in that it allows large signal transients whilst guaranteeing proper operation of the converter (switches and resonant network) and a fast response without overshoot and with minimum duration.

A4.2   ASSUMPTIONS

The circuit of the converter under consideration is that of Fig. A1.1; we will make the following assumptions:

• The switches are ideal: the switching times, the forward volt-drop and the leakage current in the off-state are negligible.

• The input and output voltages are perfectly smooth. Thus the output quantity to be controlled is the current rather than the voltage. Furthermore, if the transients are sufficiently short compared with the time constants of the input and output filters the input and output voltages can be taken as constant.

• The resonant network is perfect and lossless.

• The transformer is perfect and has a 1:1 turns ratio.

• The influence of the snubbers and of the protection circuits of the inverter and of the rectifier can be neglected.

Under these assumptions, the state plane (v_C, $j\sqrt{L/C}$) provides a simple description of the simultaneous evolution of the voltage across the capacitor, the current in the resonant network (and thus in the switches), and thus the energy stored in the resonant network. This representation applies to the steady state as well as to the transients.

A4.3   ANALYSIS

Since any trajectory in the state plane results from the elimination of time between the equations for the voltage across the capacitor and the current in the resonant network, control strategies that do not involve the time interval between two commutations are more easily studied in the state plane. Let us give the example of the control of the voltage across the capacitor or the control of the conduction time of the transistors (or the diodes) in the inverter that makes use of switching loci defined entirely by the control strategy.

The state plane analysis (Fig. A4.1) also shows very clearly the intrinsic limitations of the transient performance of the resonant converter. The energy stored in the resonant network can only increase or decrease with limited 'derivatives' which are strictly load dependent.

At the current zero crossing, the energy in the resonant network is stored in the capacitor. This energy increases more rapidly when the diodes of the inverter are not conducting. The voltage increase in a half-period is:

$$\Delta V^+ = 2(1 - q)E_1 \qquad (A4.1)$$

This quantity tends to zero as q tends to 1. By contrast, the energy decreases the more rapidly when only the diodes of the inverter are conducting. The transistors are always off and the voltage decrease in a half-period is:

$$\Delta V^- = 2(1 + q)E_1 \qquad (A4.2)$$

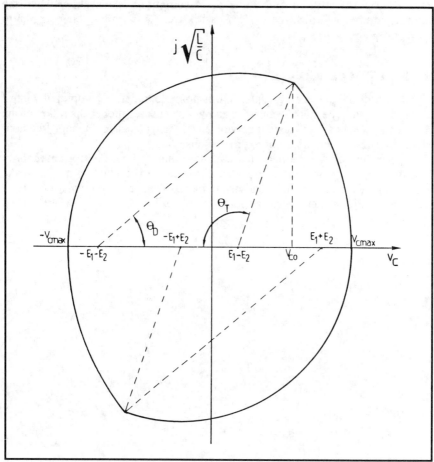

Fig. A4.1 Representation of the operation of the series resonant converter in the state plane ($f_s > f_0$)

This quantity never reaches zero and remains always greater than two. The rate of decrease of energy stored in the resonant network is thus always greater than the rate of increase.

In steady state, for each value of the control parameter there is a particular trajectory which is generally a limit cycle. This limit cycle is fully determined by the centres of the various circular arcs of which it is composed, that is to say by the input and output voltages E_1 and E_2 or by the normalized voltage $q = E_2/E_1$, and by the normalized peak voltage across the capacitor $x_M = v_{Cmax}/E_1$.

In transient conditions and under the assumptions stated above,when the control parameter varies, the system passes from one limit cycle (q,x_{M1}) to another limit cycle (q,x_{M2}).

A4.4 OPTIMAL CONTROL

In the light of the preceeding analysis, it is possible to define an optimal control consistent with large signal transients and which ensures proper operation of the converter (switches and resonant network) and yields a fast, overshoot-free non-asymptotic response with a minimum response time.

In operation above the natural frequency, this optimal control amounts to defining the state plane trajectories that correspond to conduction modes of the inverter diodes in the desired steady-state mode Thus, two commutation circles are defined (Fig. A4.2) centred at points with abscissa $(1+q)$ and $- (1+q)$ with a radius R:

$$R = x_{M2}+1+q \tag{A4.3}$$

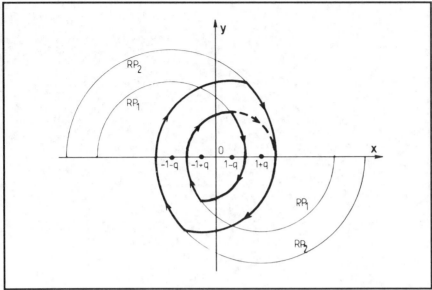

Fig. A4.2 Optimal control: definition of the commutation circles in the state plane. Each commutation circle defines a trajectory corresponding to steady state operation. The 'transient' trajectory (dotted line) is followed to change from steady state RP$_1$ to steady state RP$_2$

The turn off of the dual thyristors is initiated as soon as the distance D between the point representing the state of the resonant network and the centre of the commutation circle, which in normalized form is:

$$D = \sqrt{y^2 + \left(x + (1+q).\text{sgn}(y)\right)^2}$$ (A4.4)

is greater than radius R.

If this condition is never satisfied, the dual thyristors must similarly be turned off after a half-period of the resonant circuit, at the current zero-crossing. Thus, the turn-off control logic B is:

$$B = (D > R) \ \underline{\text{OR}} \ (\theta_T = \pi)$$ (A4.5)

In practice, a safety margin should be allowed to make sure the dual thyristors commutate properly:

$$B = (D > R) \ \underline{\text{OR}} \ \{(\theta_T > \pi/2) \ \underline{\text{AND}} \ (y < y_{min})\}$$ (A4.6)

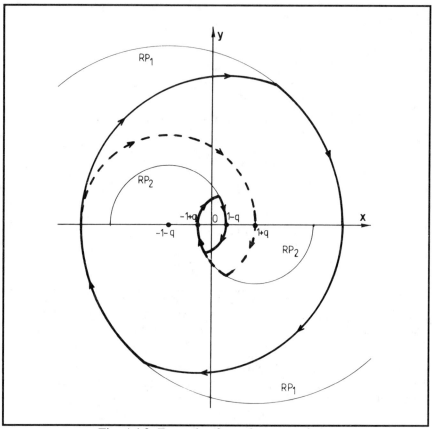

Fig. A4.3 Example of transient operation:
elimination of the mode in which energy is supplied to the resonant network

Figures A4.2, A4.3 and A4.4 show the different cases which occur when the command is changed whilst the diodes are conducting or even at the zero crossing of the current:
• Starting from the steady-state regime RP_1 in Fig. A4.2, the trajectory corresponding to the conduction of the transistors cuts the commutation circle belonging to the steady state regime RP_2 (Fig. A4.2).
• Starting from the steady-state regime RP_1 in Fig. A4.3, the energy in the resonant network is too high. As soon as the dual thyristors turn on, D is greater than the radius R and they are immediately turned off (Fig. A4.3). The mode in which the energy increases is eliminated.
• Starting from the steady-state regime RP_1 in Fig. A4.4, the energy stored in the resonant network is too low. D remains smaller than radius R and the dual thyristors stay on for half a period of the resonant network and are turned off at the zero crossing of the current. Only the mode in which the energy increases is present.

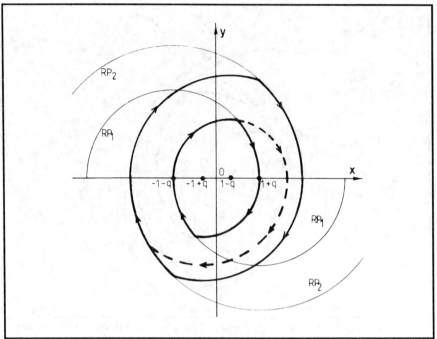

Fig. A4.4   Example of transient operation:
maximum increase of energy in the resonant network

As far as a sudden change of the output voltage is concerned, the optimal control gives a fast, overshoot-free non-asymptotic response. The short-circuit behaviour of the resonant converter is then very good (Fig. A4.5).

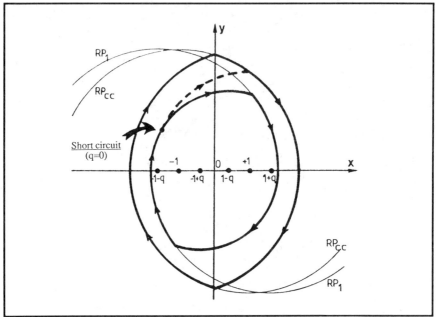

Fig. A4.5 Transient resulting from a short-circuit at the output of a series
resonant converter

A4.5 STATIC CHARACTERISTICS

Equipped with this optimal control strategy, the resonant converter exhibits the
static characteristics indicated in Fig. A4.6. Its natural behaviour is close to that
of a current source with an internal impedance independent of the load and of the
control reference R. An approximate analytic expression for the output
characteristic $q(Y_{avg})$ is:

$$Y_{avg} = 0.66R - 0.82q - 0.33 \tag{A4.7}$$

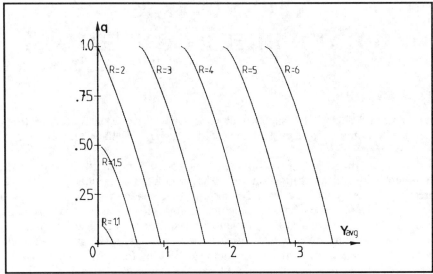

Fig. A4.6: Output characteristics of the series resonant converter with optimal control

A4.6 CONCLUSION

This optimal control strategy precisely regulates the currents and voltages in the resonant network which therefore do not overshoot during transients, especially in the case of a short-circuit.

Following a step in the reference signal R or after a sudden change of the voltage across the load, the response time is minimal and of the same order of magnitude as the natural period of the resonant network.

Under these circumstances, the series resonant converter can be essentially considered as a current generator in relation to the output filter, whose DC current can be directly adjusted by means of the reference R. With this optimal control strategy, the behaviour of the series resonant converter becomes very similar to that of the pulse-width-modulated converter.

APPENDIX A5
PHASE SHIFT CONTROL

A5.1 PRINCIPLE

With a full bridge inverter, such as that used in the series resonant inverter (Fig. A1.1), a common control method consists of varying the relative phase-shift of the two inverter legs.

However this control method is not entirely compatible with soft commutation in all operating regions. The commutation mechanisms are not precisely defined since they depend on the operating conditions.

Nevertheless, in the resonant converter of Fig. A1.1 a solution was found to allow the use of **phase shift control**. This solution consists of taking the two dual thyristors inverter legs as not one inverter supplying one resonant network but as two independent inverters supplied by the same voltage source and feeding two resonant networks in parallel with each other and seeing a common load.

This principle is shown in Fig. A5.1. The two voltage sources V_{11} and V_{12} produce unmodulated square waves with the same fixed amplitudes and frequencies, but their relative phase-shift can be adjusted by a control circuit.

Regarding each of these voltage sources in series with its resonant circuit as a current source, the power control strategy resolves to adjusting the phase-shift between these two identical current sources in the range 0 to π. The two sources are connected in parallel to a common rectifier.

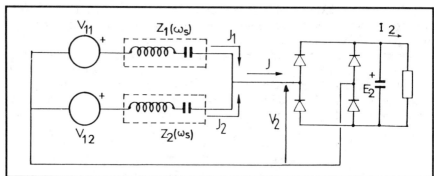

Fig. A5.1 Basic principle of phase-shift control: simplified circuit

Phase shift control, which thus gives direct control of the current drawn by the rectifier raises the possibility of a resonant converter operating at a constant switching frequency over the whole operating region.

A5.2 CHARACTERISTICS

Limiting the analysis to the fundamental components, we get from Fig. A5.1:

$$\overline{V}_{11} = \overline{V}_2 + \overline{Z}\,\overline{J}_1 \tag{A5.1}$$

$$\overline{V}_{12} = \overline{V}_2 + \overline{Z}\,\overline{J}_2 \tag{A5.2}$$

where:

$$\overline{Z}_1(\omega_s) = \overline{Z}_2(\omega_s) = \overline{Z} \tag{A5.3}$$

Let:

$$\overline{V}_0 = (\overline{V}_{11} + \overline{V}_{12})/2 \tag{A5.4}$$

From which, we can derive:

$$\overline{V}_0 = \overline{V}_2 + (\overline{Z}\,\overline{J})/2 \tag{A5.5}$$

and the diagram of Fig. A5.2.

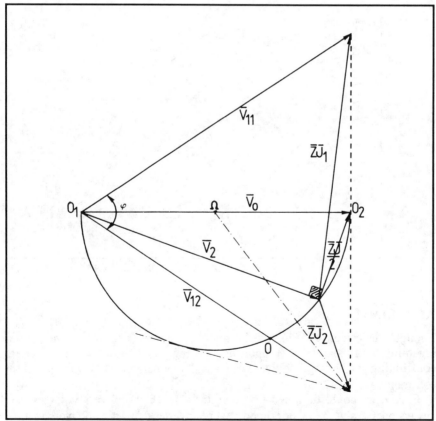

Fig. A5.2 Phase-shift control: vector diagram for the simplified circuit

The vector $\overline{V}_0$ is precisely determined by the amplitude common to $\overline{V}_{11}$ and $\overline{V}_{12}$, and their relative phase-shift φ set by the controller. Since the fundamentals of $\overline{V}_2$ and of $\overline{J}$ are always in phase, the vector $\overline{V}_0$ is the sum of two orthogonal vectors. As the load varies, the locus of the end of vector $\overline{V}_2$ thus describes a semi-circle of diameter $\overline{V}_0$.

Since $\overline{V}_2$ and $\overline{J}$ are quantities directly proportional to the output variables E_2 and I_2, the loaded characteristics (Fig. A5.3) of a series resonant converter using phase shift control can be derived from Fig. A5.2. These characteristics are ellipses whose half-axes are proportional to $\cos \varphi/2$.

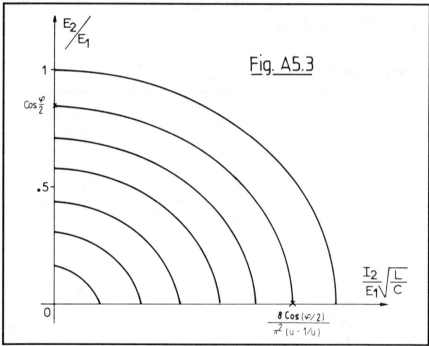

Fig. A5.3 Output characteristics (simplified circuit)

A5.3 LIMITS

Though the circuit of Fig. A5.1 is totally symmetrical, basic reasoning regarding the fundamental components and Fig. A5.2 show that the inverters constituting the voltage sources V_{11} and V_{12} operate under very different constraints.

Firstly, it should be noted that inverter (1), which delivers voltage V_{11}, always operates in the inverter mode, while inverter (2), which delivers voltage V_{12}, operates:

• in the rectifier mode when the tip of vector V_2 lies on sector OO_2,
• in the inverter mode when the tip of vector V_2 lies on sector OO_1.

Secondly, as far as inverter (2) is concerned, current J_2 passes through a minimum when the projection of $\overline{Z}\ \overline{J_2}$ passes through Ω. The difference in phase between $\overline{V}_{12}$ and $\overline{J_2}$ also reaches a minimum when the vector $\overline{Z}\ \overline{J_2}$ is tangential to the circle centred on Ω. These observations lead us to anticipate some commutation limits for inverter (2) related to the control logic of the dual thyristors and to their snubbers (c.f. appendix A3). Inverter (1), in contrast, is much less affected by commutation problems.

Finally, the current delivered by the inverter (1) is always greater than that delivered by the inverter (2), except at no-load or short-circuit when they are equal.

APPENDIX A6
CONTROLLED RECTIFICATION
AND
REVERSIBILITY

A6.1 INTRODUCTION

This appendix presents a review, using a synthesis approach of the operation of the reversible DC/DC series resonant converter using a constant switching frequency and in which power is regulated by means of controlled rectification of the current in the resonant network.

The converter, shown in Fig. A6.1, is composed of a voltage source E_1 connected to a voltage source E_2, through two voltage-source inverters CS_1 and CS_2, a resonant network and possibly a transformer.

This study is based on the analysis of trajectories in the state plane $(v_C, j\sqrt{L/C})$. It leads to precise specification of the commutation mechanisms of the switches and of the stresses applied to the active and passive components under any operating conditions.

In addition to giving a better understanding of the converter operation, it also provides the principal elements of a design.

A6.2 ASSUMPTIONS

Referring to the converter circuit shown in Fig. A6.1, we will make the following assumptions:
• The switches are ideal: the switching duration, the forward volt-drop and the off-state leakage current are negligible.
• The input and output voltages are ripple-free.
• The resonant network is lossless.
• The transformer is perfect and has a 1:1 turns ratio.
• The influence of the snubbers and of the protection circuits can be neglected.

Under these assumptions, voltages v_1 and v_2 delivered by the inverters CS_1 and CS_2 are perfect square waves. The inverter CS_1 sets the switching frequency f_s and the control signals to inverter CS_2 are phase-shifted with respect to inverter CS_1 by a phase-angle δ. Adjustment of this angle δ gives control of the power transfer.

The voltage across the resonant network is thus fixed and the duration of each mode is determined entirely by the control circuits.

The conventions and notation are those of Appendix A1, especially for the normalized quantities. In particular, remember that:
• $q = E_2/E_1$,

• u = f_s/f_0, the ratio of the switching frequency to the the natural frequency of the LC network.

Lastly, in order to compress the notation, suffix i is used to denote the value of a variable at time t_i (for instance $v_C(t_i) = v_{Ci}$).

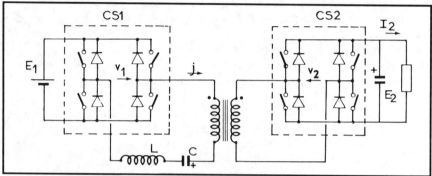

Fig. A6.1 Reversible series resonant converter: power control is achieved at a fixed frequency by controlled rectification of the current in the resonant network, which amounts to controlling the phase shift δ between the control signals for converters CS_1 and CS_2

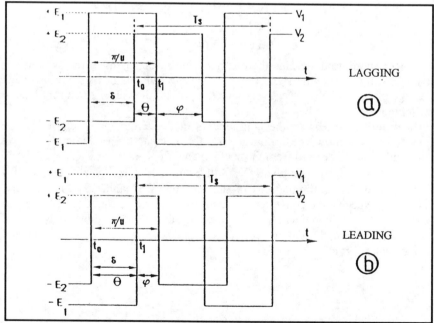

Fig. A6.2 Voltage waveforms across the resonant network

A6.3   ANALYSIS

Fig. A6.2 shows the voltages v_1 and v_2 delivered by the inverters CS_1 and CS_2. Since the value of the phase shift angle δ always lies between 0 and π/u, two cases must be considered as they do not involve the same commutation mechanisms:

• voltage v_2 lags voltage v_1 (Fig. A6.2a),
• voltage v_2 leads voltage v_1 (Fig. A6.2b).

The average value of the current in the voltage source E_2 is:

$$I_2 = \frac{2}{T_s} \int_{t_0}^{t_0+\frac{T_s}{2}} j \, dt \tag{A6.1}$$

where

$$j = C \, dv_C/dt \tag{A6.2}$$

Thus, in normalized quantities, we get:

$$Y_{2avg} = -2X_0u/\pi \tag{A6.3}$$

Similarly, the average current flowing through voltage source E_1 can be obtained:

$$Y_{1avg} = 2X_1u/\pi \tag{A6.4}$$

Furthermore, to determine the commutation mechanisms of the switches, it is helpful to examine the sign of current j at the switching times, that is to say the sign of Y_0 and Y_1 respectively which represent the values of current j at the switching times.

In order to dimension the components in the circuit, it is also of prime importance to know the amplitude of the current to be switched. These amplitudes can be derived from those of Y_0 and Y_1.

In the normalized state plane (X,Y), the points $M_0(X_0,Y_0)$ and $M_1(X_1,Y_1)$ that represent the commutations of the inverters CS2 and CS1 respectively, completely specify the converter operation. In particular, equations (A6.3) and (A6.4) show that the points M_0 and M_1 are sufficient to determine the values of the current flowing through the voltage sources, u being fixed by the design and by the switching frequency.

A6.4 STEADY-STATE OUTPUT CHARACTERISTICS

In this section, the objective is to derive the analytical equations for the coordinates of points M_0 and M_1 as functions of u, q and δ, at switching frequencies both higher or lower than the natural frequency, but limited to the continuous conduction modes (u > 0.5). This steady-state analysis leads to a family of output characteristics derived from the coordinates M_0 and M_1, which reflect the operation of the converter.

Denoting by (Fig. A6.2):
• W_1, the normalized voltage applied to the resonant circuit between t_0 and t_1,
• W_2, the normalized voltage applied to the resonant circuit between t_1 and $t_0 + T_s/2$.

Let

$$\theta = \omega_0(t_1 - t_0)$$

$$\varphi = \pi/u - \theta$$

Denoting by Z_0 and Z_1 the mappings of points M_0 and M_1 in the complex plane gives:

$$Z_1 - W_1 = (Z_0 - W_1)e^{-j\theta} \tag{A6.5}$$

$$Z_1 - W_2 = (-Z_0 - W_2)e^{j\varphi} \tag{A6.6}$$

from which it can be deduced that:

$$X_0 = \frac{W_1 - W_2}{2}\left(1 - \frac{\cos\left(\frac{\pi}{2u} - \theta\right)}{\cos\left(\frac{\pi}{2u}\right)}\right) \tag{A6.7}$$

$$Y_0 = \frac{W_1 - W_2}{2}\frac{\sin\left(\varphi - \frac{\pi}{2u}\right)}{\cos\left(\frac{\pi}{2u}\right)} - \frac{W_1 + W_2}{2}\tan\left(\frac{\pi}{2u}\right) \tag{A6.8}$$

$$Y_1 = \frac{W_1 + W_2}{2}\frac{\sin\left(\theta - \frac{\pi}{2u}\right)}{\cos\left(\frac{\pi}{2u}\right)} + \frac{W_1 - W_2}{2}\tan\left(\frac{\pi}{2u}\right) \tag{A6.9}$$

According to the principle of conservation of energy we have:

$$X_1 = -qX_0 \tag{A6.10}$$

211

A6.4.1 Voltage v_2 lags v_1 (Fig. A6.2a)

Substituting for W_1 and W_2 and taking:

$$\delta = \varphi \qquad \text{(A6.11)}$$

gives:

$$X_0 = 1 - \frac{\cos\left(\delta - \frac{\pi}{2u}\right)}{\cos\left(\frac{\pi}{2u}\right)} \qquad \text{(A6.12)}$$

$$Y_0 = \frac{\sin\left(\delta - \frac{\pi}{2u}\right)}{\cos\left(\frac{\pi}{2u}\right)} + q \tan\left(\frac{\pi}{2u}\right) \qquad \text{(A6.13)}$$

$$Y_1 = q \frac{\sin\left(\delta - \frac{\pi}{2u}\right)}{\cos\left(\frac{\pi}{2u}\right)} + \tan\left(\frac{\pi}{2u}\right) \qquad \text{(A6.14)}$$

A6.4.2 Voltage v_2 leads v_1 (Fig. A6.2a).

Substituting for W_1 and W_2 and taking:

$$d = q \qquad \text{(A6.15)}$$

gives:

$$X_0 = -1 + \frac{\cos\left(\delta - \frac{\pi}{2u}\right)}{\cos\left(\frac{\pi}{2u}\right)} \qquad \text{(A6.16)}$$

$$Y_0 = \frac{\sin\left(\delta - \frac{\pi}{2u}\right)}{\cos\left(\frac{\pi}{2u}\right)} + q \tan\left(\frac{\pi}{2u}\right) \qquad \text{(A6.17)}$$

$$Y_1 = -q \frac{\sin\left(\delta - \frac{\pi}{2u}\right)}{\cos\left(\frac{\pi}{2u}\right)} - \tan\left(\frac{\pi}{2u}\right) \qquad \text{(A6.18)}$$

Examining these different equations reveals that for a constant frequency of operation:
• The amplitude of X_0 only depends on the phase shift δ, and by virtue of equation (A6.3), the output DC current is independent of the output voltage. In other words, the converter behaves as a perfect current source controlled by the angle δ.
• On the other hand, the sign of X_0, and thus the direction of power flow, depends on the sign of the phase shift (either leading or lagging) of voltage v_2 with respect to voltage v_1.
• Figure A6.3 shows the output current variation versus the angle δ, for different values of parameter u. For a phase shift δ equal to zero or to π/u, no power is exchanged between the voltage sources E_1 and E_2. The output current reaches its peak when the phase shift δ between the voltages v_2 and v_1 equals $\pi/2u$, leading or lagging. Two different ranges of phase shift δ give the maximum output current swing: either δ less than $\pi/2u$, leading or lagging, or δ greater than $\pi/2u$, leading or lagging. Thus the first step in the design is to choose one of these ranges. As a result of this choice, the stresses suffered by the various components in the circuit are very different.

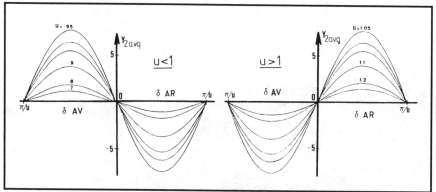

Fig. A6.3 Average output current versus control angle δ

• The value of Y_0 depends on the phase shift δ and the output voltage q, but is independent of whether the phase-shift is leading or lagging. This also means that the commutation mechanisms in inverter CS_2 only depend on the power transferred, but not on the direction of power flow.
• The values of Y_1 from Equations (A6.14) and (A6.18) are opposite but it should be noted that they correspond to different commutation processes in inverter CS_1 (Fig. A6.2). Consequently, the commutation mechanisms of inverter CS_1 also depend only on the power transferred and not on the direction of flow.

A6.5 ANALYSIS OF THE COMMUTATION

Having shown that the converter behaves as a current source controlled by the phase shift δ, it now becomes sensible to clarify the commutation mechanisms of the switches in inverters CS_1 and CS_2 under any operating condition. With frequency u and the output voltage q held constant, a study of the loci described by points M_0 and M_1 as δ varies results in the definition of the various operating regions within which the commutation mechanisms are known exactly.

The analytic equation of each of these loci is obtained by elimination of the parameter δ in the expressions for X_0 and of Y_0 on the one hand, and for X_1 and of Y_1, on the other hand.

A6.5.1 Voltage v_2 lags voltage v_1

$$\left(X_0-1\right)^2+\left(Y_0-q\,\tan\!\left(\frac{\pi}{2u}\right)\right)^2=\frac{1}{\cos^2\!\left(\frac{\pi}{2u}\right)}$$

(A6.19)

$$\left(X_1+q\right)^2+\left(Y_1-\tan\!\left(\frac{\pi}{2u}\right)\right)^2=\frac{q^2}{\cos^2\!\left(\frac{\pi}{2u}\right)}$$

(A6.20)

A6.5.2 Voltage v_2 leads voltage v_1

$$\left(X_0+1\right)^2+\left(Y_0-q\,\tan\!\left(\frac{\pi}{2u}\right)\right)^2=\frac{1}{\cos^2\!\left(\frac{\pi}{2u}\right)}$$

(A6.21)

$$\left(X_1-q\right)^2+\left(Y_1-\tan\!\left(\frac{\pi}{2u}\right)\right)^2=\frac{q^2}{\cos^2\!\left(\frac{\pi}{2u}\right)}$$

(A6.22)

These loci are plotted in Figs A6.4 and A6.5 where u is less than 1 and u is greater than 1, respectively. The symmetry of these figures confirms the comments made in Section A6.4.

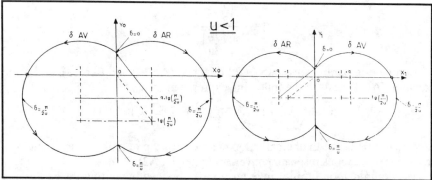

Fig. A6.4 u<1: locus of the points of commutation (M_0 for converter CS_2 and M_1 for converter CS_1 in the state plane)

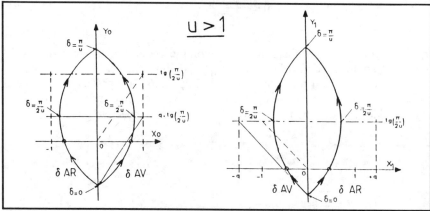

Fig. A6.5 u>1: locus of the points of commutation (M_0 for converter CS_2 and M_1 for converter CS_1) in the state plane

When δ is less than π/2u, leading or lagging, these loci cross the x-axis for certain values of the frequency and output voltage. This change of sign of the commutated current leads to a change in the commutation mechanisms of the switches. Thus, with switches operating in the soft commutation mode, there is a commutation limit that prevents the output current from varying over the full range.

On the other hand, when δ is greater than π/2u leading or lagging, these loci never cross the x-axis. Thus there is no commutation problem in this phase shift range. If u is greater or less than one all the switches are turn-off or turn-on controlled respectively.

The commutation limit of inverter CS_2 is obtained by putting the expression for Y_0 (A6.13) equal to zero, hence:

$$q= \frac{\sin\left(\frac{\pi}{2u} - \delta\right)}{\sin\left(\frac{\pi}{2u}\right)}$$

(A6.23)

Eliminating parameter δ between Equations (A6.12) and (A6.23), and using equation (A6.3) leads to the following expression:

$$q^2 \sin^2\left(\frac{\pi}{2u}\right) + \left(1 + \frac{\pi}{2u} Y_{2avg}\right)^2 \cos^2\left(\frac{\pi}{2u}\right) = 1$$

(A6.24)

Again, this is the mathematical expression for the output characteristic of a non-reversible, series resonant converter (Appendix A2).

The commutation limit of inverter CS_1 is obtained by putting the expression for Y_1 (A6.14) equal to zero, thus:

$$q= \frac{\sin\left(\frac{\pi}{2u}\right)}{\sin\left(\frac{\pi}{2u} - \delta\right)}$$

(A6.25)

and, as is inherent from the converter topology, we note that equations A6.23 and A6.25 are the inverses of each other.

These commutation limits are shown in Fig. A6.6 for u greater than one and u less than one. The commutation limits of inverters CS_1 and CS_2 have their concave sides towards the top or towards the bottom, respectively.

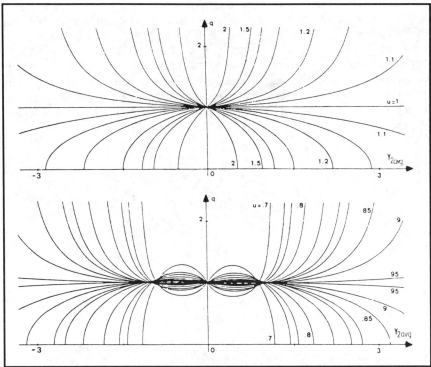

Fig. A6.6: Commutation boundaries for converters CS_1 and CS_2 in the plane $q(Y_{2avg})$

A6.6 COMPONENT STRESSES

The analysis so far carried out satisfactorily describes the operation of the reversible resonant converter from the point of view of the load. However, it does not make apparent the stresses applied to the various components. The value of the current in the switches and in the resonant network, as well as the voltage across the capacitor or the inductor, are strictly necessary to design these components. To this end, extra curves giving the maximum amplitude of the current flowing through the resonant network are plotted on the $q(Y_{2avg})$ plane.

Figure A6.7 shows a set of curves, each of them made up of all the steady state points with current of the same amplitude flowing in the resonant network. Each curve of this set is plotted for a certain value of u and with the normalized quantity y_{max} as a parameter:

$$y_{max} = j_{max}/E_1\sqrt{C/L} \qquad (A6.26)$$

217

Moreover, in most cases, the resonant current can be considered as quasi-sinusoidal. The evaluation of the voltage across the resonant network is then straightforward.

From Fig. A6.7, it can be seen that the ratings are closely related to the operational mode chosen. For a given operating point (Y_{2avg}, q), the current in the resonant network with leading or lagging δ greater than $\pi/2u$ is always very much greater than that obtained with δ less than $\pi/2u$. The compromise between the ratings and the operating limits due to the commutation of inverters CS_1 and CS_2 should be noted.

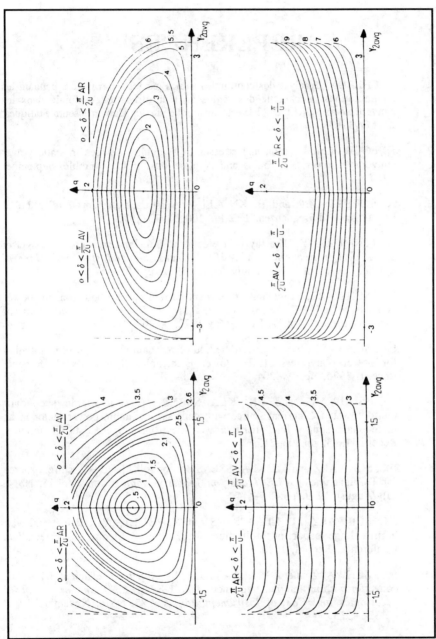

Fig. A6.7 Loci in the plane q(Y_{2avg}) of the points corresponding to a fixed peak current in the resonant network. These loci correspond to different values of the normalized quantity $Y_{max} = j_{max}/E_1\sqrt{C/L}$

219

REFERENCES

1   Y. CHERON, 'Analyse des contraintes subies par les interrupteurs pendant les commutations. Recherche des règles de leur utilisation optimale dans les convertisseurs.' GRECO 22 Electrotechnique sur les Convertisseurs Statiques, 21 May1987.

2   H. FOCH, 'Commutation and stresses of switching devices in static power converters.'Power Electronics and Applications, EPE, Grenoble, September 1987.

3   A. BOEHRINGER and H. KNÖLL, 'Transistorschalter im Bereich hoher Leistungen und Frequenzen.' *ETZ* Bd. 100, No 13, 1979.

4   W. MacMURRAY, 'The thyristor electronic transformer: a power converter using a high frequency link.' *IEEE Trans. on Industry and General Applications*, Vol.7, No 4, July/August 1971, pp.451~457.

5   F.C. SCHWARZ, 'A method of resonant current pulse modulation for power converters.' *IEEE Trans. on Industrial Electronics and Control Instrumentation*, Vol.17, No 3, May 1970, pp.209~221.

6   F.C. SCHWARZ, 'An improved method of resonant current pulse modulation for power converters.' IEEE Power Electronics Specialists Conference, Record, 1975, pp.194~204.

7   F.C. SCHWARZ and J.B. KLAASSENS, 'A controllable secondary multikilowatt DC current source with constant maximum power factor in its three phase supply line.' IEEE Power Electronics Specialists Conference, Record, 1975, pp.205~215.

8   F.C. SCHWARZ and J.B. KLAASSENS, 'A controllable 45~kW current source for DC machines.' *IEEE Trans. on Industry Applications*, Vol.15, No 4, July/August 1979, pp.437~444.

9   F.C. SCHWARZ, J.B. KLAASSENS and W. PETIET, 'An efficient 600 watt high voltage capacitor multiplier.' IEEE Power Electronics Specialists Conference, Record, 1980, pp.316~325.

10   F.C. SCHWARZ and J.B. KLAASSENS, 'A 95-percent efficient 1-kW DC converter with an internal frequency of 50 KHz.' *IEEE Trans. on Industrial Electronics and Control Instrumentation*, Vol.25, No 4, November 1978, pp.326~333.

11   F.C. SCHWARZ and J.B. KLAASSENS, 'A reversible smooth current source with momentary internal response for nondissipative control of multikilowatt DC machines.' *IEEE Trans. on Power Apparatus and Systems*, Vol.100, No 6, June 1981, pp.3008~3016.

220

12   F.C. SCHWARZ and W.L.F.H. MOIZE De CHATELEUX, 'A multi-kilowatt polyphase AC/DC converter with reversible power flow and without passive low frequency filters.' *IEEE Trans. on Industrial Electronics and Control Instrumentation*, Vol.28, No 4, November 1981, pp. 273~281.

13   F.C. SCHWARZ, 'A doubled-sided cycloconverter.' *IEEE Trans. on Industrial Electronics and Control Instrumentation*, Vol.28, No 4, November 1981, pp. 282~291.

14   F.C. SCHWARZ and J.B. KLAASSENS, 'A radar power supply without voltage droop.' IEEE Power Electronics Specialists Conference, Record, 1983, pp. 230~241.

15   J.B. KLAASSENS, 'DC to AC series resonant converter system with high internal frequency generating synthesized waveforms for multikilowatt power levels.' IEEE Power Electronics Specialists Conference, Record, 1984, pp.99~110.

16   J.B. KLAASSENS, 'DC/AC series resonant converter system with high internal frequency generating multiphase AC-waveforms for multikilowatt power levels.' IEEE Power Electronics Specialists Conference, Record, 1985, pp.291~301.

17   SANNIER and GARCIA, 'Convertisseur Continu-Continu à étage intermédiaire moyenne fréquence.' Graduation dissertation, 1976.

18   H. FOCH and J. ROUX, 'Convertisseurs statiques d'énergie électrique à semiconducteurs' ANVAR patent, France No 7832428, RFA No P29452457, GB No 7939217, USA No 093106, Italy No 83487A/79.

19   H. FOCH, P. MARTY and J. ROUX, 'Use of duality rules in the conception of transistorised converters' PCI Proceedings, Munich, 1980, pp.4B3 1~11.

20   Y. CHERON, 'Application des règles de la dualité à la conception de nouveaux convertisseurs à transistors de puissance. Synthèse du Thyristor-dual. Domaine d'application.'Thèse de Docteur- Ingénieur, INP Toulouse, 1982.

21   Y. CHERON, H. FOCH and J. ROUX, 'Etude d'une nouvelle utilisation des transistors de puissance dans les convertisseurs haute tension à fréquence élevée.' *Revue de Physique Appliquée*, June1981, pp.333~342.

22   Y. CHERON, H. FOCH and Y. MOPTY, 'A new medium frequency inverter combining asymmetrical thyristors and power transistors.' PCI Proceedings, Geneva, 1982, pp.347~359.

23   Y. CHERON, H. FOCH and Y. MOPTY, 'Un convertisseur moyenne fréquence à thyristors et à transistors de puissance.' Le Transistor de Puissance dans la Conversion d'Energie, Thomson-CSF, March 1983.

24   Y. MOPTY, 'Méthode de synthèse automatique des convertisseurs statiques. Aplication à la recherche de nouveaux convertiseurs.'Thèse de Docteur-Ingénieur, INP Toulouse, 1982.

25   B. ESCAUT and P. MARTY, 'Introduction à l'étude des structures des convertisseurs statiques. La commutation dans les convertisseurs statiques.' *Electronique Industrielle* No 56 (1983), 58 (1983), 60 (1983) and 64 (1984).

26   L. GYUGYI and B.R. PELLY, 'Static power frequency changers. Theory, performance, and application.' J. Wiley & Sons, 1976.

27   T. WIDODO, 'Etude des convertisseurs statiques de fréquence. Synthèse des structures — commandes — analyses.'Thèse de Docteur-Ingénieur, INP Toulouse, 1981.

28   A. PERIN, 'Contribution à l'étude des convertisseurs directs de fréquence à transistors de puissance.'Thèse de Docteur- Ingénieur, INP Toulouse, 1984.

29   B. CHAUVEAU, 'Etude la commande d'un onduleur alimentant une charge non linéaire à partir d'une source impédante: Optimisation de la distorsion harmonique de la tension de sortie.'Thèse de Doctorat de l'I.N.P. Toulouse, 1987.

30   A. COURTEIX, Y. CHERON, H. FOCH and M. METZ, 'Application à un convertisseur continu-alternatif complexe d'une méthode systématique de synthèse.' *Revue de Physique Appliquée* No 21, June1986.

31   A. COURTEIX, 'Contribution à la synthèse systématique des convertisseurs statiques à conversion indirecte. Application à la conversion continu basse tension/alternatif basse fréquence.'Thèse de Doctorat de l'I.N.P. Toulouse, 1986.

32   J. LAGASSE, 'Etude des circuits électriques.' Eyrolles, 1965.

33   N. BALABANIAN and T.A. BICKART, 'Electrical network theory.' J. Wiley & Sons, 1969.

34   S. CÜK, 'General topological properties of switching structures.' IEEE Power Electronics Specialists Conference, Record, June 1979, pp.109~130.

35   R. SEVERNS, 'Switchmode converter topologies — Make them work for you!' Intersil Application Bulletin A035, 1980.

36   F. BRICHANT, 'L'ondistor.' Dunod, 1970.

37   Y. CHERON and H. FOCH, 'Utilisation de la résonance dans les alimentations de puissance.' *Electronique de Puissance*, No 2, 1983, pp.45~48.

38   N. MAPHAM, 'An SCR inverter with good regulation and sine-wave output.' *IEEE Trans. on Industry and General Applications*, Vol.3, No 2, March/April 1967, pp.176~187.

39   G.N. REVANKAR and S.A. GADAG, 'Analysis of high frequency inverter circuit.' *IEEE Trans. on Industrial Electronics and Control Instrumentation*, Vol.20, No 3, August 1973, pp.178~182.

References

40   G.N. REVANKAR and S.A. GADAG, 'A high frequency bridge inverter with series parallel compensated load.' *IEEE Trans. on Industrial Electronics and Control Instrumentation*, Vol.21, No 1, February 1974, pp.18~21.

41   R. KASTURI, 'An analysis of series inverter circuits.' *IEEE Trans. on Industrial Electronics and Control Instrumentation*, Vol.22, No 4, November 1975, pp.515~519.

42   J. CHEN and R. BONERT, 'Load independent AC/DC power supply for high frequencies with sine-wave output.' *IEEE Trans. on Industry Applications*, Vol.19, No 2, March/April 1983, pp.223~227.

43   S. BOYER, H. FOCH, J. ROUX and M. METZ, 'Chopper and PWM inverter using GTO's in dual thyristor operation.' Power Electronics and Applications, EPE, Grenoble, September 1987.

44   P. TRZCINSKI, 'Etude et réalisation d'un convertiseur Continu-Alternatif à modulation de largeur d'impulsion.'Thèse de Docteur-Ingénieur, Orsay, 1980.

45   H. FOCH and J. ROUX, 'Convertisseur 750V-1kW à transistors de puissance destiné à la charge de batterie d'accumulateurs.' Contract report MATRA, 1979.

46   R.L. STEIGERWALD and R.E. TOMPKINS, 'A comparison of high- frequency link schemes for interfacing a DC source to a utility grid.' IAS Annual Meeting, 1982, pp.759~766.

47   R.L. STEIGERWALD, A. FERRARO and F. G. TURNBULL, 'Application of Power Transistors to Residential and Intermediate Rating Photovoltaic Array Power Conditionners.' *IEEE Trans. on Industry Applications*, Vol.19, No 2, March/April 1983, pp.254~267.

48   P. CABALEIRO CORTIZO, 'Techniques de mise en série des transistors de puissance pour la moyenne et haute tension.'Thèse de Docteur-Ingénieur, INP Toulouse, 1984.

49   B.D. BEDFORD and R.G. HOFT, 'Principles of inverters.' J. Wiley & Sons, 1964.

50   H. FOCH, ' Les convertisseurs statiques à commutation forcée. Etudes — Synthèses — Simulations.'Thèse de Docteur ès-Sciences Physiques, Université Paul Sabatier, Toulouse, 1974.

51   M. METZ, 'Simulation analogique globale des systèmes à interrupteurs semiconducteurs par interfaces spécialisées (Procédé CASSIS). Caractérisation en régime permanent et régime transitoire des circuits de commutation forcée des convertisseurs statiques à thyristors.'Thèse de Docteur ès-Sciences, INP Toulouse, 1985.

52   P.M. ESPELAGE and B.K. BOSE, 'High Frequency Link Power Conversion.' *IEEE Trans. on Industry Applications*, Vol.13, No 5, September/October 1977, pp.387~394.

References

53   L. GYUGYI and F. CIBULKA, 'The High Frequency Base Converter - A New Approach to Static High Power Conversion.' *IEEE Trans. on Industry Applications*, Vol.15, No 4, July/August 1979, pp.420~429.

54   I. BARBI, ' Etude d'onduleurs autoadaptatifs destinés à l'alimentation de machines asynchrones.'Thèse de Docteur- Ingénieur, INP Toulouse, 1979.

55   F. POGNANT, 'Circuits de commutation forcée autoadaptatifs à thyristors. Méthode d'étude des régimes transitoires. Application au réglage de vitesse rhéostatique d'une machine asynchrone.'Thèse de Docteur-Ingénieur, INP Toulouse, 1980.

56   K.H. LIU, R. ORUGANTI and F.C. LEE, 'Resonant Switches - Topologies and Characteristics.' IEEE Power Electronics Specialists Conference, Record, 1985, pp.106~16.

57   F.C. LEE, 'Zero-voltage switching techniques in DC-DC converter circuits.' High Frequency Power Conversion Conference, Washington, April 1988.

58   T.A. MEYNARD, Y. CHERON and H. FOCH, 'Generalization of the resonant switch concept. Structures and performances.' High Frequency Power Conversion Conference, Washington, April 1988.

59   T.A. MEYNARD, Y. CHERON and H. FOCH, 'Generalization of the resonant switch concept. Structures and performances.'Power Electronics and Applications, EPE, Grenoble, September 1987.

60   T.A. MEYNARD, Y. CHERON and H. FOCH, 'Généralisation du concept d'interrupteur résonnant. Structures et performances.' *Revue Générale de l'Electricité*, No 2, Février 1988.

61   T.A. MEYNARD, Y. CHERON and H. FOCH, 'Soft switchings in DC/DC converters : reduction of switching losses and of EMI generation.' High Frequency Power Conversion Conference, San Diego, 2~4 May 1988.

62   T.A. MEYNARD, 'Commutation douce appliquée aux alimentations à découpage'Thèse de Doctorat de l'I.N.P. Toulouse, 1988.

63   J.J. LE ROUX, B. CANDELIER and A. KAZDAGHLI, 'Application of the asymmetrical thyristor switching power supply (15 kHz/10 kW).' PCI Proceedings, Geneva, 1982, pp.360~371.

64   F. BORDRY, Y. CHERON, H. FOCH and M. METZ, 'Analyse des méthodes d'étude et de simulation des convertiseurs. Application à un convertisseur à résonance.' Journées SEE, l'Electronique de Puissance du Futur, Grenoble, June 1985.

65   R. ORUGANTI and F.C. LEE, 'Resonant Power Processors: Part I - State Plane Analysis.' IAS Annual Meeting, 1984, pp. 860~867.

References

66   Y. CHERON, H. FOCH and J. SALESSES, 'Study of a resonant converter using
      power transistors in a 25-kW   X-Rays tube power supply.' IEEE Power
      Electronics Specialists Conference, ESA Proceedings, 1985, pp.295~306.

67   R. ORUGANTI and F.C. LEE, ' State Plane Analysis of Parallel Resonant
      Converters.' IEEE Power Electronics Specialists Conference, Record, 1985,
      pp.56~73.

68   V. NGUYEN, J. DHYANCHAND and P. THOLLOT, 'Steady-state and small-
      signal analysis of series-resonant converter: a novel graphical approach.' IAS
      Annual Meeting, 1986, pp. 692~701.

69   D. CHAMBERS, ' Designing High Power SCR Resonant Converters for Very
      High Frequency Operation.' Powercon 9, Record, F2 pp.1~12.

70   R.C. COLE, 'A Gated Resonant Inverter Power Processor for Pulsed Loads.'
      IEEE Power Electronics Specialists Conference, Record, 1981, pp.312~326.

71   B. HENNEVIN, S. VOLUT and A. DUPAQUIER, 'Study and realization of a
      37.5 kW double resonance converter using GTO thyristors.' Power Electronics
      and Applications, EPE, Grenoble, September 1987, pp.371~376.

72   R.J. KING and T.A. STUART, 'A Normalized Model for the Half-Bridge Series
      Resonant Converter.' *IEEE Trans. on Aerospace and Electronic Systems*,
      Vol.17, No 2, March 1981, pp.190~198.

73   V. VORPERIAN and S. CUK, 'A Complete DC Analysis of the Series Resonant
      Converter.' IEEE Power Electronics Specialists Conference, Record, 1982,
      pp.85~100.

74   R.J. KING and T.A. STUART, 'Transformer Induced Instability of the Series
      Resonant Converter.' *IEEE Trans. on Aerospace and Electronic Systems*,
      Vol.19, No 3, May 1983, pp.474~482.

75   R.J. KING and T.A. STUART, 'Inherent Overload Protection for the Series
      Resonant Converter.' *IEEE Trans. on Aerospace and Electronic Systems*,
      Vol.19, No 6, November 1983, pp.820~829.

76   R.J. KING and T.A. STUART, 'Modelling the Full Bridge Series Resonant
      Power Converter.' *IEEE Trans. on Aerospace and Electronic Systems*, Vol.18,
      No 4, July 1984, pp.449~459.

77   P.D. ZIOGAS, V.T. RANGANATHAN and V.R. STEFANOVIC, 'A Four
      Quadrant Current Regulated Converter with a High Frequency Link.' *IEEE
      Trans. on Industry Applications*, Vol.18, No 5, September/October 1982,
      pp.499~505.

78   V.T. RANGANATHAN, P.D. ZIOGAS and V.R. STEFANOVIC, 'A Regulated
      DC-DC Voltage Source Converter Using a High Frequency Link.' *IEEE Trans.
      on Industry Applications*, Vol.18, No 3, May/June 1983, pp.279~287.

79   V.T. RANGANATHAN, P.D. ZIOGAS and V.R. STEFANOVIC, 'A DC-AC
      Power Conversion Technique Using Twin Resonant High Frequency Links.'
      IEEE Trans. on Industry Applications, Vol.19, No 3, May/June 1983,
      pp.393~400.

80   V.T. RANGANATHAN, P.D. ZIOGAS and V.R. STEFANOVIC, 'Performance
      Characteristics of High Frequency Links Under Forward and Regenerative
      Power Flow Conditions.' IAS Annual Meeting, 1983, pp.831~839.

81   R.L. STEIGERWALD, 'High Frequency Resonant Transistor DC-DC
      Converters.' *IEEE Trans. on Industrial Electronics*, Vol.31, No 2, May 1984,
      pp.181~191.

82   V. VORPERIAN and S. CUK, 'Small Signal Analysis of Resonant
      Converters.' IEEE Power Electronics Specialists Conference, Record, 1983,
      pp.269~282.

83   J.P. FERRIEUX, 'Modèlisation des convertisseurs Continu-Continu à
      découpage.'Thèse de 3ème cycle, I.N.P. Grenoble, 1984.

84   V. VORPERIAN, 'High-Q Approximations in the Small Signal Analysis of
      Resonant Converters.' IEEE Power Electronics Specialists Conference,
      Record, 1985, pp.707~715.

85   K. AL HADDAD, Y. CHERON, H. FOCH and V. RAJAGOPALAN, 'Static and
      dynamic analysis of a series resonant converter operating above its resonant
      frequency.' SATECH'86 Proceedings, Boston, 1986, pp. 55~68.

86   K. AL HADDAD, Y. CHERON, H. FOCH and V. RAJAGOPALAN, 'Etude des
      performances statiques et dynamiques d'un convertisseur à résonance
      fonctionnant à une fréquence supérieure à sa fréquence de résonance.' *Canadian
      Journal of Electrical and Computer Engineering*, Vol. 12, No 4, 1987.

87   K. AL HADDAD, 'Etude des différentes stratégies de contrôle d'un
      convertisseur à résonance. Application à l'alimentation haute tension d'un
      LASER CO_2.' Memoir ofThèse de Doctorat de l'I.N.P.Toulouse, February 1988.

88   R. ORUGANTI and F.C. LEE, 'Resonant Power Processors: Part II — Methods
      of Control.' IAS Annual Meeting, 1984 , pp. 868~878.

89   Y. CHERON, H. FOCH and J. ROUX, 'Power transfer control methods in high
      frequency resonant converters.' PCI Proceedings, Munich, 1986, pp.92~103.

90   Y. CHERON, H. FOCH and J. ROUX, 'Convertisseurs à résonance: Méthodes
      de contrôle du transfert de puissance.' *Electronique de Puissance* No 16,
      September 1986.

91   Y. CHERON, P. JACOB and J. SALESSES, 'Dispositif statique de réglage des
      échanges d'énergie entre des systèmes électriques générateur et/ou récepteur.'
      ANVAR patents: France No 8511291, Europe No 86.201231.7, USA No
      4.717.998.

References

92   P. SAVARY, M. NAKAOKA and T. MARUHASHI, 'Resonant Vector Control Base High Frequency Inverter.' IEEE Power Electronics Specialists Conference, Record, 1985, pp.204~213.

93   P. SAVARY, M. NAKAOKA and T. MARUHASHI, 'Novel type of high-frequency link inverter for photovoltaic residential applications.' *IEE Proceedings*, Vol. 133, Pt. B, No 4, July 1986.

94   P. JACOB, 'Groupes de secours statiques à hautes performances. Application des principes de la résonance.'Thèse de Doctorat de l'I.N.P. Toulouse, 1986.

95   D. DIXNEUF, 'Etude d'un variateur de vitesse à résonance pour machine asynchrone triphasée 15 kVA, 440 V, 60 Hz.'Thèse de Doctorat de l'I.N.P. Toulouse, 1988.

96   Y. CHERON, D. DIXNEUF, 'Dispositif perfectionné pour le réglage des échanges d'énergie entre des systèmes électriques générateur et/ou récepteur.' ANVAR patent extension: France No 87.15659.

97   P. PROUDLOCK, 'The medium power auxiliary magnet power supplies for LEP, can switched-mode meet the challenge?' PCI Proceedings, Genève, 1982, pp.297~309.

98   M. BOIDIN, H. FOCH, Y. CHERON and P. PROUDLOCK, 'The design, construction and evaluation of a new generation high frequency 40-kW DC converter. ' PCI Proceedings, Paris, 1984, pp.124~133.

99   M. BOIDIN, H. FOCH, Y. CHERON and P. PROUDLOCK, 'Etude, construction et évaluation d'un convertisseur 40 kW/20 kHz à résonance. ' *Electronique de Puissance* No 9, April 1985.

100   P. JACOB, G. DUNAND-FRARE and P. SEROT, 'Convertisseurs à haut rendement et faible EMI-RFI. ' *Electronique de puissance* No 13, Février 1986.

101   P. JACOB, 'Changeurs de fréquence à résonance haute fréquence. ' *Electronique de puissance* No 24, December 1987.

102   B. CHAUVEAU, Y. CHERON, H. FOCH and M. BINET 'Etude de la compensation des non-linéarités de la source et de la charge dans un onduleur PWM' *Revue Canadienne de Génie Electrique et Informatique*, 1987.

103   B. CHAUVEAU, Y. CHERON, H. FOCH and M. BINET 'On the control of a PWM inverter supplied by a fluctuating voltage source and feeding a non-linear load' Power Electronics and Applications, EPE, Grenoble, September 1987.

104   B. CHAUVEAU, Y. CHERON and H. FOCH 'Etude de la commande d'un onduleur monophasé sur charge non linéaire. Perspectives' D.R.E.T.contract final report No 841139, June 1986.

105   H. FOCH et l'équipe Convertisseurs Statiques 'Méthodes d'études des convertisseurs statiques.' Mentor.April 1988

Index